Conceptualization and Treatment Planning for Effective Helping

Conceptualization and Treatment Planning for Effective Helping

BARBARA F. OKUN
Northeastern University

KAREN L. SUYEMOTO
University of Massachusetts

Australia • Brazil • Japan • Korea • Mexico • Singapore • Spain • United Kingdom • United States

Conceptualization and Treatment Planning for Effective Helping
Barbara F. Okun and Karen L. Suyemoto

Publisher: Jon-David Hague

Acquiring Sponsoring Editor: Seth Dobrin

Assistant Editor: Naomi K. Dreyer

Assistant Editor: Suzanna R. Kincaid

Associate Media Editor: Elizabeth Momb

Marketing Program Manager: Tami Strang

Art and Cover Direction, Production Management, and Composition: PreMediaGlobal

Manufacturing Planner: Judy Inouye

Rights Acquisition Specialist: Dean Dauphinais

Cover Image: Pakhnyushcha/Shutterstock

© 2013 Brooks/Cole, Cengage Learning

ALL RIGHTS RESERVED. No part of this work covered by the copyright herein may be reproduced, transmitted, stored, or used in any form or by any means graphic, electronic, or mechanical, including but not limited to photocopying, recording, scanning, digitizing, taping, Web distribution, information networks, or information storage and retrieval systems, except as permitted under Section 107 or 108 of the 1976 United States Copyright Act, without the prior written permission of the publisher.

> For product information and technology assistance, contact us at **Cengage Learning Customer & Sales Support, 1-800-354-9706**.
>
> For permission to use material from this text or product, submit all requests online at **www.cengage.com/permissions**.
> Further permissions questions can be e-mailed to **permissionrequest@cengage.com**.

Library of Congress Control Number: 2011946211

ISBN-13: 978-1-133-31405-9

ISBN-10: 1-133-31405-8

Brooks/Cole
20 Davis Drive
Belmont, CA 94002-3098
USA

Cengage Learning is a leading provider of customized learning solutions with office locations around the globe, including Singapore, the United Kingdom, Australia, Mexico, Brazil, and Japan. Locate your local office at **www.cengage.com/global**.

Cengage Learning products are represented in Canada by Nelson Education, Ltd.

To learn more about Brooks/Cole, visit **www.cengage.com/brookscole**.

Purchase any of our products at your local college store or at our preferred online store **www.cengagebrain.com**.

Printed in the United States of America
1 2 3 4 5 6 7 16 15 14 13 12

DEDICATION

Dedicated in memoriam to Donna Raymer and James Meyer, Jr. with thanks for all they gave to us, some small part of which we pass on to others here.

Contents

About the Authors — xi
Preface — xii
Ancillaries — xiv
Acknowledgements — xv

INTRODUCTION — XVI

Overview of the Text — xvii
   Section I — xvii
   Section II — xviii

CHAPTER ONE: EXPLORING CONCEPTUALIZATION — 1

Foundational Understandings — 2
What Is Conceptualization? — 4
   Conceptualization Relates to but Is More than Theoretical Orientation — 5
   Conceptualization Is a Relational Process — 6
   Conceptualization Is an Integral Part of Evidence Based Practice — 7
   Conceptualization Is a Continuous Process — 10
Introducing the Case of Nancy — 10
   Referral Source — 10
   Background — 11

Presenting Concerns and Strengths	12
Verbal and Nonverbal Behavior	12
Attitude towards Therapy	13
Conceptualizing Nancy	13
Summary	14
References	14

CHAPTER TWO: EXPLORING INFLUENCES OF PERSONAL WORLDVIEWS — 17

Why Is Theoretical Orientation Important?	17
Understanding Your Beliefs about Change	19
Influences on Your Ideas about Change	21
Cultural Influences on Understanding Clients: Beliefs about People and Worldview	21
Power, Privilege, and Inequity	22
Power and Privilege Related to the Role of the Therapist	24
Beliefs about Health and Pathology: The Social Context of Psychotherapy	26
Criteria for Evaluating Health and Pathology	27
Influences on Your Conceptualization of Health and Pathology	30
Introducing Dimensions of Change	33
Summary	37
References	37

CHAPTER THREE: CONCEPTUALIZING CLIENTS IN CONTEXTS — 39

Exploring Contexts	39
Exploring Family Contexts	44
Exploring Sociocultural and Sociostructural Systems	46
Dimensions Related to Contextual Understandings	52
Context Location of the Problem	53
Focus of Change	58
Summary	64
References	64

CHAPTER FOUR: CONCEPTUALIZING THERAPEUTIC RELATIONSHIPS — 67

Understanding Your Relational Style	68
Directive versus Non-directive	68
Structuring Process	70
Level of Activeness	71
Confrontativeness	72

viii CONTENTS

Relation of Dimensions of Relational Style to Established Theories	74
Relational Style Dimensions and the Role of the Therapist	75
Understanding the Role of the Relationship in Contributing to Change	76
The Working Alliance	77
Significance of the Therapeutic Relationship	79
Real-Unreal Relationship	82
Process Emphasis	87
Interactions of Relationships and Location of Knowledge	90
Pulling It All Together	91
Summary	95
References	95

CHAPTER FIVE: DILEMMAS IN EFFECTIVE HELPING — 97

Fit of Therapist and Client	98
Boundary Dilemmas	108
Disclosure	110
Boundary Issues Related to Managed Care Systems	115
Social Justice and Cultural Sensitivity	116
Clients' Internalized Oppression	118
Clients' Bias and Discrimination Related to Therapist Minority Status	120
Consultation and Supervision	121
Summary	122
References	122

CHAPTER SIX: BEGINNING CONCEPTUALIZATION: GATHERING AND INTEGRATING INFORMATION — 125

Beginning the Process of Case Conceptualization: Gathering Information	125
Primary Sources of Information	127
Secondary Sources of Information	142
Organizing and Integrating Information (Pre-conceptualization)	145
Connecting Information to Theory: Moving towards Conceptualization	146
Summary	148
References	149

CHAPTER SEVEN: CONCEPTUALIZATION, TREATMENT PLANNING, AND DIAGNOSIS — 151

Integrating Information and Moving towards Conceptualization with Nancy	151
Interactions of Information Gathering with Early Conceptualization and Treatment Planning	156
Initial Treatment Planning during Information Gathering with Nancy	157
Conceptualization and Treatment Planning	158
Therapists' Intervention Preferences	158

Treatment Planning: A Journey, Not a Destination	160
Developing Goals	162
Choosing Strategies for Treatment Goals	169
Diagnosis	171
Formalizing Conceptualizations and Treatment Plans	175
Summary	176
References	176

CHAPTER EIGHT: ITERATIVE CONCEPTUALIZATION AND TREATMENT PLANNING — 179

Continual Conceptualization with Nancy	181
Continual Conceptualization and Integrating New Information	183
Revisiting Organizing Information for Continual Conceptualization	184
Testing Hypotheses from Prior Information and Conceptualizations	184
Encountering Surprises: The Process of Iterative Conceptualization	188
Iterative Conceptualization from Start to Finish: The Case of Juan	194
Initial Information	194
Initial Conceptualization and Treatment	197
Iterative Conceptualization and Treatment	199
Iterative Conceptualization and Modifying Diagnoses	203
Iterative Conceptualization, Consultation, and Adjunct Treatments	204
Ongoing Records and Clinical Notes	205
Interactions with Systems of Care	206
Summary	206
Reference	207

CHAPTER NINE: CONTINUOUS EVALUATION AND TERMINATION — 209

Continuous Evaluation of Clients and Their Progress in Therapy	210
Continuous Evaluation of Our Effectiveness as Therapists	211
Evaluating Client Progress	212
Choosing to Continue with New Goals or to End Therapy	218
Termination	219
Unplanned Termination	221
Finances and Termination	223
Planned Termination	223
Termination with Nancy	227
Follow-up and Returning Clients	228
Follow-up	228
Returning Clients	229
Final Thoughts	229
References	231

Appendix A
A Brief Review and Application of Established Theories   **232**

Review of Theoretical Approaches   233
    Psychodynamic Theories   233
    Cognitive-Behavioral Theories   236
    Existential–Humanistic   238
    Systems–Ecological Model   240
    Constructivist Theories   242
    Liberation Perspectives: Feminist/Multicultural   246
References and Recommended Readings   248

Appendix B
Exploring Your Experiences with Culture, Power, and Privilege   **250**

Brief Review of Exploring Your Ethnic Culture   250
Brief Review of Exploring Your Power and Privilege   253
References and Resources   255

Appendix C
Nancy's Genogram   **257**

Index   **258**

About the Authors

Barbara F. Okun, Ph.D., is a Professor of Counseling Psychology at Northeastern University and a clinical instructor at Harvard Medical School. She is Training Director of the Counseling/School Doctoral program at Northeastern and teaches and supervises clinical courses. She is the author of numerous books and articles and is currently preparing the eighth edition of *Effective Helping: Interviewing and Counseling Techniques*. Dr. Okun trains mental health professionals internationally and is currently working on a project to develop a doctoral program in Zambia.

Karen L. Suyemoto, Ph.D., is an Associate Professor in Psychology and Asian American Studies at the University of Massachusetts, Boston. She has taught theory and practicum courses at undergraduate and graduate levels. She has provided consultation and training for therapists both locally and nationally, with particular focus on developing cultural sensitivity. Her research and teaching focus generally on issues related to social justice and anti-racist therapy, research, and education, particularly in relation to racialized identities, meanings of race and ethnicity, and contributing to social justice from both oppressed and privileged spaces.

Preface

Over the years, as we have taught and supervised graduate (and undergraduate) students in developing their understanding and skills as counselors, we have been faced with the challenge of helping mental health trainees "find themselves" as psychological helpers and in relation to specific clients at a given time. In our teaching we have, of course, used many helpful books and resources to teach theory and skills. And one of us (BFO) has solved one part of the teaching challenge by writing a book (*Effective Helping: Interviewing and Counseling Techniques*) to address the development of the knowledge and skills necessary for developing the basic foundation of an effective helping relationship: a working alliance and an introductory knowledge of helping strategies. However, we have found it difficult to find the book that would follow the introductory book, a book that would develop the knowledge and awareness and skills needed by professional helpers as they work over time with clients who have varying degrees of difficulty. Professional therapists need training that goes beyond developing a working alliance, identifying problems, and understanding the content of theory and the types of strategies available. So we decided to tackle the challenge of teaching trainees *how* (not what) to think about clients and how to apply that understanding to treatment planning and choices.

This book addresses that challenge. The chapters within aim to help trainees develop awareness of why they choose what they do, in relation to theoretical orientation *and* in relation to the choices they make with a specific client in a particular moment. Our goal is to help trainees learn to integrate the many interacting pieces of the helping process (including theoretical orientation, the person of the therapist, the person of the client, the contexts that affect the clients, the therapy relationship, the context of the therapy, the skills and resources available, and so forth) into a conceptualization that will foster efficacy in creating positive change. We simultaneously hope that trainees will develop an understanding of how they approach this integration, so they may continuously develop their conceptualization skills throughout their own professional development. In sum, this book was written to assist mental health trainees to integrate their self awareness, theoretical orientation, and understanding of human behavior from a developmental and ecological perspective with an assessment of clients in contexts, in order to form a client-centered conceptualization for effective helping.

We understand conceptualization as a dynamic process of contextual hypothetical thinking in clinical situations. We see case conceptualization as a therapist's story about the client and about the client's own story. And we see both of these stories as always evolving as the helping relationship develops. Thus, conceptualizing cases is continuous, changing, and involves the helper's understanding of self and personal theories of change. Case conceptualization is what guides treatment planning, which is also a continuous process. Case conceptualization and treatment planning are, therefore, a circular process involving the space between the helper and the client and other contextual circumstances and changes.

The book is divided into two sections. The chapters in Section I aim at helping trainees understand and develop their theoretical orientation, by exploring areas of values and worldview and introducing dimensions that can be useful to therapists in developing their own integrative theoretical orientations. These chapters also explore how theoretical orientation relates to case conceptualization. More specifically, Section I focuses on (1) defining and exploring the complex meaning of conceptualization; (2) exploring trainees' understanding of how change happens and effects of their worldviews and values on this understanding; (3) defining and exploring the contexts that affect clients and that must be considered to develop a wholistic conceptualization; (4) exploring important dimensions of the therapeutic relationship and their effects on case conceptualization and treatment planning, and (5) examining some common dilemmas encountered by novice therapists, particularly those related to issues of fit between therapist and client, boundaries, and social justice.

In Section I, we introduce a framework of Dimensions of Change, which we use as an organizational and integrative tool throughout the section. This framework aims to help trainees develop an integrative theoretical framework congruent with their worldview through differentiating dimensions of therapy related to the attitudes and action of the therapist, the relationship between the client and the therapist, and the focus for creating change. We use this framework to help trainees consider how their own values, thinking, and worldview affect their approach to understanding pathology, thinking about clients and their contexts, and the way in which they use the therapeutic relationship. By considering these as dimensions, rather than choices or stances, we aim to help trainees consider how, when, and why they may choose to position themselves differently on a given dimension in order to best help a specific client.

Section II focuses on the process of conceptualization and treatment planning, integrating the knowledge and awareness gained through Section I. In this section, we aim to elucidate how an integrative orientation is applied to case conceptualization, goal setting, and treatment planning; how good conceptualization is always an interaction between the particular client and the therapist's theoretical understanding and skills; and how conceptualization is a continuous process. Section II is more practically oriented, taking trainees through the process of developing and revising conceptualization and treatment planning. We conclude with a consideration of evaluation and termination.

Overall, the goals of this book are to: 1) help trainees develop their own theoretical orientation; 2) explore the process of conceptualization and the influence of the trainee's theoretical orientation on the conceptualization process; 3) identify the values, worldviews, and contexts that affect the therapist, the client, and the therapeutic relationship in order to ensure these are used for effective conceptualization and intervention; 4) help the novice therapist develop skills for gathering and organizing information and knowledge and to integrate this knowledge with self and other awareness into case conceptualization for evidence based practice; and 5) examine and develop skills for the process of iterative conceptualization.

Throughout the book we provide experiential exercises. We introduce the case of Nancy in Chapter 1 and utilize this case throughout the text to illustrate the progression through assessment, conceptualization, intervention, re-conceptualization, and termination. However, we also provide many additional case examples in both the text and exercises in order to illustrate how clients vary and to examine the interaction of contextual, person, and relational variables that affect conceptualization and treatment planning. We also provide additional material in the Appendices (Appendix A: A Brief Review and Application of Established Theories and Appendix B: Exploring Your Experiences with Culture, Power, and Privilege) that we expect will be review for most readers, but may be helpful in reminding trainees and consolidating previous learning.

Ancillaries

An Instructor's Manual is available for this text. This manual provides a number of resources including:

- A discussion of general pedagogical approaches for teaching case conceptualization, and intervention planning in didactic and practicum courses.
- A chapter-by-chapter outline and summary of major points.
- An instructor's guide for discussion facilitation related to the content and exercises in each chapter.
- Suggestions for further study, including additional readings and activities.
- Essay questions for assessing students' learning.

Acknowledgements

This book could not have been written without our graduate students and our clients from whom we have learned so much over the years. We particularly would like to thank the following students who provided feedback on the drafts and contributed to the development of our thinking during the actual writing process: Jessica Graham, Kelly Graling, Sarah Gray, Fernanda Lucchese, Frances Martinez-Pedraza, Sarah Schwartz, Akhila Venkatachalam, Speshal Walker. We also want to acknowledge Monica Torreiro-Casals, a doctoral student at Northeastern University, who worked on the literature searches and references. Thanks also to Lizabeth Roemer and Beth Boyd for sharing their expertise in response to questions. And thanks to Eric Parker for providing us with sustenance, time, and patience during our writing retreats.

We also want to thank our editor Seth Dobrin and other members of the Cengage team who have made this book possible, Julie Martinez and Pooja Khurana. In addition, the reviewers for both the first and final drafts of this manuscript have been fantastic! Their thoughtful, thorough suggestions have been incorporated and we truly appreciate their efforts.

Introduction

Psychotherapists are helping professionals in formal helping relationships, where the primary goal is to help the client. In these formal relationships, the roles of the helper and the client are clear and usually structured by multiple helping contexts such as the treatment context, professional organizations, legal systems, and third-party payers. In addition to understanding themselves, helping professionals need to understand the different ways people change, learn, and develop socially, physically, cognitively and emotionally, what determines health and illness, and how to integrate this information with the unique experiences and contexts of the individual client.

Our goal in this textbook is for you to be able to formulate your own case conceptualization with awareness of what you draw from established theories, how and why you make those choices, and how your understanding affects treatment decisions. We view conceptualization as a continuous relational process and, while we do not ascribe to any one theoretical perspective, we believe that each of the major models of psychotherapy has something of value to contribute to our understanding of clients and our efforts to help them. We are starting from an assumption that readers of this book are near, but not at, the very beginning of their professional journeys. We are therefore assuming that the reader of this book has:

- *Some basic familiarity with theories of psychological development and psychotherapy.* While we describe the relation of traditional theories to dimensions of understanding change and provide an overview of these theories in Appendix A, we are assuming that this overview is review and that the reader has greater familiarity with these theories than we provide in this text.
- *Some general self-awareness.* This includes some self-knowledge about your own likes and dislikes, your reasons for wanting to be a professional helper, your strengths and challenges, and some reflection on the kinds of experiences that have shaped you into the person you are today.

- *Some awareness of cultural differences and the meaning of power and systems of privilege, power dynamics in relationships, and different experiences and reactions to having or not having personal, interpersonal, and sociostructural power.* We are assuming that you have some understanding of your own cultured experience and power and privilege, related to ethnicity and race as well as to other issues such as gender, social class, sexual orientation, and other socially important statuses. We are also assuming you have some knowledge about people different from you, and the influences culture, power, and privilege have on different groups of people and on interactions. Appendix B provides a review of basic multicultural and power issues.

- *Some training and skills in developing positive relationships between helpers and clients.* Other texts (for example, *Effective Helping: Interviewing and Counseling Techniques*) focus primarily on the meaning and importance of, and skills contributing to, a positive working alliance. We are assuming that you have had some training in developing the basic characteristics of an effective helper at the relational stage, including: self-awareness, cultural awareness, honesty, congruence, ability to communicate well, and ethical integrity.

OVERVIEW OF THE TEXT

The book is divided into nine chapters in two sections, with three appendices. References are provided at the end of each chapter, and your instructor may provide additional readings and resources. Throughout the book, we use case examples and experiential exercises to enable the reader to develop self-awareness and better understanding of clients in unique contexts and the therapy process.

Section I

Section I (Chapters 1–5) is focused on exploring your understanding about change, which influences your theoretical orientation which, in turn, serves as the foundation of your case conceptualization. While case conceptualization is primarily focused on understanding the client, we believe that your understanding is inherently shaped by your own values, worldviews, and preferences. A therapist's own values, feelings, and responses can be invaluable sources of information *or* can be sources of blindspots and biases that can be detrimental to thorough understanding and to facilitating change in psychotherapy. The difference between whether the person of the therapist is a facilitating or impeding factor in the change process is related to how much the therapist is aware of his or her own views and how these views are affecting the case conceptualization and the process of therapy. The chapters in Section I aim at helping trainees understand and develop their theoretical orientation and understanding of influences on their conceptualizations, by exploring areas of values and worldview and introducing dimensions that can be useful to therapists in developing their own integrative theoretical orientations and conceptualizing across theoretical orientations. In addition, Chapter 5 explores common dilemmas for novice therapists that relate to your developing theoretical orientation, case conceptualizations, and treatment planning.

In Chapter 1, *Exploring Conceptualization*, we define conceptualization and clarify the relationship between your theoretical orientation and your clinical conceptualizations. We present the goals of the book overall and explore conceptualization as a continuous relational process. We also discuss the relationship of conceptualization to Empirically Supported Therapy (EST), Empirically Validated Therapy (EVT), and Evidence Based Practice (EBP). We introduce the case of Nancy, which will be used throughout the text.

Chapter 2, *Exploring Influences of Personal Worldviews*, focuses on the values and worldviews that influence your choices of theoretical orientation and your approach to case conceptualization. We discuss different theoretical views on health and pathology from an ecological perspective. We also introduce dimensions of change and the approach to change that relate to your theoretical orientation and case conceptualization.

Chapter 3, *Conceptualizing Clients in Contexts*, is focused on clarifying your beliefs about people, contexts, and relationships. We explore the multiple ways that contexts, particularly family and sociocultural/sociostructural systems, affect your own and your clients' experiences and consider how a relative emphasis within these systems will affect your case conceptualization. As we explore these contexts, we consider differing emphases on affect, cognition, or behavior as well as the past/present continuum.

Chapter 4, *Conceptualizing Therapeutic Relationships*, considers your understanding of the therapeutic relationship and its influence on change. We review different dimensions related to your relational style, and the nature and centrality of the therapeutic relationship, and consider how your preference within these dimensions might relate to your theoretical orientation and your case conceptualization.

Chapter 5, *Dilemmas in Effective Helping*, presents common clinical dilemmas by exploring the complexities that may arise in the "gray" areas in ethical issues, such as the fit between therapist and client, issues related to boundaries, and dilemmas related to therapy and social justice including clients' internalized oppression and discrimination towards therapists from minority statuses. How you make meaning of these dilemmas is shaped by your theoretical orientation, the therapeutic relationship, and your conceptualization and continuous evaluation of the therapeutic relationship and the treatment.

Section II

Section II focuses on the process of conceptualization and its relation to treatment planning. Within this section, we aim to integrate the self-awareness you have developed from your progress through Section I with skills in gathering and integrating information, and using this information to develop a wholistic understanding of the client in contexts, with particular attention to understanding the ways that the presenting problems are developed or maintained and the associated best ways to intervene. In this section, we aim to elucidate how an integrative orientation is applied to case conceptualization, goal setting, and treatment planning; how good conceptualization is always an interaction between the particular client and the therapist's theoretical understanding and skills; and how conceptualization is a continuous process. We conclude with a chapter focused on evaluating progress and issues related to termination.

Chapter 6, *Beginning Conceptualization: Gathering and Integrating Information*, focuses on how you observe, collect, and organize the information from which your initial conceptualization develops. This includes how you hear and make meaning of (assess) your clients' stories—the presentation, self-descriptions, and perceptions of their problems. We also focus on how you integrate information from other providers, family, and perhaps previous records into your firsthand observations and experiences of the client.

In Chapter 7, *Conceptualization, Treatment Planning, and Diagnosis*, the focus is on how your understanding of the client translates into case conceptualization, which integrates theoretical orientation, diagnosis, and treatment planning. We explore the ways that we begin case conceptualization and treatment planning even as we are initially gathering information. We also explore how your own preferences and experiences might affect your approach to conceptualization and interventions. We address the development of goals for therapy and the ways that goal setting may be affected by the treatment context. Finally, we discuss the relation of diagnosis and conceptualization, discussing some of the controversies and the benefits of the Diagnostic Systematic Manual (DSM) and how manifestation and experience of a particular disorder can vary across clients. We use the case of Nancy to illustrate the issues we discuss.

In Chapter 8, *Iterative Conceptualization and Treatment Planning*, we focus on the ways that case conceptualization is a continuous process as we integrate new information as treatment progresses. Thus, new information activates the circular loop of reassessing earlier conceptualization and, thus, revising goals and treatment plans. We continue to utilize the case of Nancy to illustrate this continuing process and introduce a new case, Juan, whom we follow throughout a therapy to further illustrate the continuous process.

In Chapter 9, *Continuous Evaluation and Termination*, we explore the process of continuous evaluation of the client, the therapist, the therapeutic relationship, and the treatment. We also discuss the process and types of termination, including making referrals, transferring clients, and dealing with returning clients after time. We emphasize lifetime professional development and how we always need to consider changing individual, family, and societal challenges and attitudes.

1

Exploring Conceptualization

Textbooks for training therapists[1] frequently focus on developing relational skills for creating and maintaining a working alliance and initial interviewing; overviewing established theoretical orientations; or developing intervention skills and techniques. The first two types of texts are wonderful resources for those who are just beginning their training, while the third type is most helpful to those who are seeking to develop tools for helping beyond common factors. However, these texts do not address a major developmental task for training therapists: how to develop or choose your own theoretical orientation and use this for wholistic case conceptualization. Case conceptualization is the foundation of treatment planning, of making choices between the multitudes of intervention techniques to best meet the needs of a specific client.

This book focuses on developing skills in conceptualization, addressing questions such as: How do your own worldview, values, and contexts affect your theoretical orientation and case conceptualization? How do you integrate the established theories you have studied and use yourself as a vehicle for change as the helping process evolves? How do you understand the client

---

1. By "therapist" we mean someone who is or will be a professional helper within the area of mental health. Throughout the book, we use "counselor" and "therapist" interchangeably although we realize that different disciplines within applied mental health may use different language, such as counselor, therapist, clinician, and so on. By our choice of language, we are not meaning to specify any particular discipline or subdiscipline. We also use "student" and "trainee" interchangeably.

in context? How do you organize and integrate the information you need to do so? What questions must you ask yourself to continually review that understanding and revise goals and objectives? How do you carry on with the process of change involving you and your client and the evolving subjective experiences between the two of you after the initial sessions focused on gathering background information and establishing a working alliance? As your own values, attitudes, and beliefs influence your interpretation of established theories, what do you think shapes your espoused and real theoretical identity and its application to specific cases through your case conceptualization?

This book addresses the central questions: How do you understand the process of change? and How do you apply this understanding to a particular client, with sensitivity to the unique experiences and contexts of that client and the unique relationship between you? Books and articles already exist on how to organize and write a case conceptualization (for example, Stevens & Morris, 1995), or how to conceptualize a case from particular theoretical orientations (for example, Berman, 2010), or for a particular diagnosis (for example, Hersen & Porzelius, 2002). This book is more about developing your own integrative theoretical orientation and its application to specific clients: considering what you will draw from established theoretical orientations (and why) and how you will use this information in the development of your case conceptualization. It is also about considering the issues that will influence your foci for a given client within the range of theoretical understandings that speak to you and exploring the ways that your theoretical orientation and case conceptualization will affect your treatment planning. This book is also not a techniques or strategies book. It is about *how you think about and understand change*, how you integrate information, how you relate to clients, and how you plan interventions to facilitate change. It is also about how these processes are continuous, and continuously changing.

FOUNDATIONAL UNDERSTANDINGS

Our approach to this book rests on some foundational understandings about people and about learning. Before we explore an introduction to conceptualization and theoretical orientation, we want to situate ourselves and this text within these understandings.

First, our approach to case conceptualization, to therapy, and to understanding people in general is founded within an ecological model of human development (see Chapter 3 for a deeper discussion). The ecological model, as a practice perspective (Germain, 1979), considers individuals as embedded in family systems which, in turn, are embedded in the communities in which they reside, such as neighborhoods, schools, work, and religious institutions, which are embedded in larger sociocultural systems such as government, societal, and world systems. Interacting with all of these systems are critical variables such as gender, race, ethnicity, sexual orientation, cultural identities, class, and immigrant generation. Our emphasis on an ecological context relates to our belief that people are very complex and are constantly interacting with their diverse contexts. This means that we also believe that it is impossible to capture the fullness of lived experience in a single (or even several) variables. In addition, we relate this ecological approach to our belief that good conceptualization is not just problem focused, but aims to understand the whole client in multiple contexts.

Second, our foundational perspective emphasizes the diversity of human experience. While we think that there are some basic experiences that characterize people in general (for example, biological experiences such as breathing), we think that there are very few experiences that are experienced or expressed universally, particularly those that are not inherently tied to our physiology; (for example, breathing may be common across people, but even in this we are aware of influences of culture, such as the ways in which Buddhist monks experience breathing). There are always exceptions or differences, and these are not inherently pathological just because they are differences (see Chapter 3 and the discussion of different bases of health and pathology). Thus, we have a more constructivistic epistemological philosophy. This does not mean we do not value the approach of describing shared experiences of groups of people. It does mean that we attend very strongly to the difference between a modal experience of a group, and the individual experience, as well as to the variables and influences on what is shared *and* what is different. This means that we prioritize recognizing that not everyone changes in the same way and that not everyone facilitates change in the same way.

Third, we believe that some of the best learning (particularly about complex or relational processes like psychotherapy) is a mix of knowledge, reflection, and experience and that having an integrated explanatory story about change is helpful. Thus, it is frequently not enough to know what technique is best for a particular problem, how to do a technique, and how it is "supposed" to work. It is important to know how you (specifically) understand how this technique will contribute to change more generally, why you choose to use it or not (in general or at a particular time), and why some techniques are more attractive to you. Our approach to this book is to try and foster your understanding of your story through providing knowledge and opportunities for reflection and experiential learning.

The chapters contain many exercises that encourage you to more thoroughly reflect, discuss, explore, apply, and integrate the issues we discuss. We believe that having an integrative understanding that links concepts and behaviors in a web of meaning is more helpful than having isolated bits of knowledge. Developing an integrative understanding and becoming aware of how you use this in conceptualization is different than developing an eclectic approach that primarily matches technique to problems and does not actively link conceptualization of problems and treatment planning into an overall picture. We do not necessarily believe that one story or web of meaning is the right one, but that the process of developing the story and the web is, itself, helpful.

Finally, we are starting from an assumption that novice therapists and therapy trainees have positive motivations and good intentions in their desire to help others. Perhaps, this is so obvious that it could go unsaid, but we feel that it is important to make this explicit because our experience is that trainees sometimes feel criticized or discouraged by feedback that is intended to be helpful. For example, moments where teachers (or authors) point out areas that have been unexamined or areas where students or trainees have less knowledge or inaccurate knowledge are sometimes experienced by students as negative judgments. We want to be explicit that we see these areas as normal developmental challenges for trainees evolving into good therapists. We expect that trainees will have some areas where they are less knowledgeable, and that their good intentions will motivate them to address these areas. Now that we have described our basic assumptions, we turn now to exploring the meaning of conceptualization.

WHAT IS CONCEPTUALIZATION?

Conceptualization is an active process of integrating and interpreting (deriving, inferring) information with a goal of deeper understanding that moves beyond summary of instances or facts. Sperry (2005) describes case conceptualization as "a method and process of summarizing seemingly diverse case information into a brief, coherent statement or 'map' that elucidates the client's basic pattern of behavior" (p. 354). He describes case conceptualization as having three components:

1. A diagnostic formulation, which is primarily descriptive and focuses on understanding the client's psychological presentation. A diagnostic formulation is primarily related to the present and focused on the "what" question.
2. A clinical formulation, which focuses on explaining how the client's psychological presentation developed and what contributes to its maintenance. The clinical formulation is the aspect of case conceptualization that most links to theoretical orientation. The clinical formulation is related to both past and present and focused on the "why"; it provides the connection between the diagnostic formulation and the treatment formulation.
3. A treatment formulation, which focuses on choosing and describing interventions and considering their effects. The treatment formulation is more future-focused on the "how," that is, "How can change be created?"

While we find this definition and the specific components of case conceptualization helpful, we are cognizant of the many questions we have received from trainees over the years, which can be summarized as: "I have an idea of what case conceptualization is, but *how do I do it*?" Our view is that when trainees ask this question, they are asking not only "How is it done?" but also "How do *I* do it?" That is, we believe that they are working to develop an approach to case conceptualization that reflects their own particular understandings of people and change, and their own (frequently integrative) theoretical orientation. Conceptualization is your story about the client's story. As such, it reflects your beliefs and your knowledge about people and contexts, problem development and change; your own process of contributing to change as a therapist (theoretical orientation); *and* your understanding about this unique client and his or her understanding of his or her experience.

Conceptualization Relates to but Is More than Theoretical Orientation

One of the things most helpful about Sperry's (2005) description is that it identifies both the relation and the distinction between conceptualization and theoretical orientation. The clinical formation part of conceptualization is inherently related to theoretical orientation. Theoretical orientation is your understanding of how people in general develop and maintain psychological health or difficulties. Change in psychotherapy is essentially a deliberate attempt to develop and maintain particular thoughts, feelings, and behaviors; theoretical orientation describes how you think people in general change and how therapy can best contribute to that change. Because conceptualization is clearly related to theoretical orientation, one aspect of developing conceptualization skills is developing or becoming aware of one's own theoretical understandings of change. This can, itself, be challenging for therapists in training, as they (like most therapists) are frequently seeking an integration of understandings from multiple traditions of theoretical orientation. Thus, *one goal of this book is to help therapists in training develop their own theoretical orientation by exploring what aspects of different theoretical orientations resonate for them and why.*

Theoretical orientation and conceptualization are not, however, the same thing. Theoretical orientation is your general, usual, or preferred understanding of psychological development and change. For most therapists, their theoretical orientation is the approach to understanding that comes most "naturally" or immediately in most cases. It is your story about people generally. Theoretical orientation represents our attempt to articulate the "how" of development, the process of change for people both generally and in therapy.

In contrast, conceptualization is the application of theoretical orientation to a specific client with particular issues, strengths, and contexts. While your theoretical orientation may develop over your lifetime of experience as a therapist, it is unlikely to change dramatically in relatively short time spans such as months or days once it is developed. But your conceptualization should be continually changing, both from client to client and as a continuous process in relation to one specific client. The ways in which you apply (or modify) your theoretical orientation is not constant, even if your base understanding and preference is more stable. Conceptualization is also *more than* theoretical orientation, as it integrates specific knowledge and information about this particular client and about effective strategies for particular problems or populations.

The theoretical orientation that you use as a foundation for conceptualization will usually be the one with which you are most comfortable and the one you apply to people generally; in contrast, your conceptualization of one client versus another will not be the same. Clients' unique presentations will relate to different aspects of your theoretical orientation. They may even challenge you to shift out of your usual preferred theoretical orientation because a particular presenting problem is best conceptualized or treated from a different understanding. For example, a therapist who describes himself as psychodynamic may conceptualize a client with a flying phobia through a cognitive-behavioral approach, integrating his knowledge about the research on effective treatments for phobias. Your theoretical orientation does not determine how you think in every instance, or determine that you are not capable of thinking otherwise. Theoretical orientation, as filtered through our own being and experience, provides guidelines for testing out our hypotheses about who our clients are, how and why they came to therapy, with what motivation and unique styles, and how best we can help them to solve their problems and live more satisfying lives.

Case conceptualization is, therefore, applied theoretical orientation tailored to the specific client and the client's contexts, aimed at understanding the issues that bring the client to therapy and indicating the most effective interventions. What we learn from the client about his or her history, presenting problems, and perceived strengths is filtered through our theoretical orientation, our understanding of human behavior, development, and change. Conceptualization is your story about this particular client. Thus, an additional task for novice therapists is to integrate this personal client centered "conceptualization" with formal/theoretical conceptualization, tailoring to the particularities of clients. A *second major goal of this book is to explore the process of conceptualization and the influence of theoretical orientation on the varying moments or processes that are conceptualization.*

Conceptualization Is a Relational Process

Therapy is inherently a relational endeavor. Some mental health practitioners adhere more to a medical model, which is based on symptoms leading to a diagnosis and treatment to address these specific symptoms, but we believe that "symptoms" are embedded in ecological contexts and that psychotherapy itself is one of these contexts. Techniques and strategies used for intervention are always affected by the relationship. Our emphasis on the inherent relational nature of therapy relates to our understanding that the distinction that is sometimes made between common factors and specific factors (Wampold, 2001) may be less helpful to novice therapists. Common factors refer to those that cut across different theoretical orientations and strategic approaches; specific factors refer to those associated with a particular theory or approach. Common factors might include the therapeutic relationship, communication of empathy, or focus on client's agenda whereas specific factors might refer to a specific strategy such as relaxation therapy or interpretation of transference.

Specific factors or mechanisms of change work or do not work within the relational context of therapy. They frequently take part of their meaning and efficacy in creating change from the relational context of therapy and the associated common factors. Clients do not distinguish between common and specific factors. Furthermore, the relational context means that it is just as necessary to understand the meaning that the client is ascribing to an intervention as it is to understand your own intention in choosing the intervention. What you intend may or may not be what they get from the intervention. And understanding the client's meaning will be more helpful to contributing to an iterative conceptualization process than understanding only your intention.

We therefore also understand conceptualization as a relational process. By this, we mean that the person of the therapist and the person of the client affect the conceptualization, particularly the clinical and treatment formulations, as well as the intervention process itself. The values and experiences and contexts of the therapist shape not only how he or she thinks about psychological development (theoretical orientation), but also how he or she relates to others. In addition, the client's own understanding of him or herself will affect the therapist's conceptualization. It is difficult to develop treatment goals and engage the client in techniques or change processes if these are based in a conceptualization that is at odds with how the client understands him or herself (although changing this understanding may then be a treatment goal, if your conceptualization indicates that this is part of the difficulty). Because conceptualization is your story about the client's story, it is inherently co-constructed. *A third major goal of this book is to identify the values, worldviews, and contexts that affect the therapist, the client, and the relationship between them, and to explore how these relate to conceptualization.*

Conceptualization Is an Integral Part of Evidence Based Practice

Research on psychotherapy aims to explore the best and most effective ways for therapy to contribute to change. One kind of research (Norcross, 2002) attempts to identify common factors, as discussed above, that contribute to positive outcomes in therapy. For example, this research is the basis of the consensus in the field regarding the importance of the therapeutic alliance. Other research specifically aims to examine the effectiveness of specific therapeutic approaches for specific problems. In the past two decades, much attention has been paid to identifying the best treatments for specific disorders to ensure accountability and quality of psychotherapy services. The three major terms used within this area of study are: 1) Evidence Based Therapy (or Evidence Based Practice in Psychology-EBPP); 2) Empirically Supported Therapy (EST); and 3) Empirically Validated Therapy (EVT).

The latter two terms (EST and EVT) seem to be used interchangeably in the literature and aim to establish that studies meeting rigorous scientific standards have demonstrated that some interventions are most effective for given disorders and should, therefore, be chosen for particular presenting problems (Antony & Barlow, 2010; Castonguay & Beutler, 2006; Norcross, 2002). This approach assumes that effectiveness is best established through studies that meet rigorous scientific standards such as replicated group and single design experiments. Typically, empirically supported therapy studies compare the outcomes of the target treatment to those of alternative treatments. Treatment manuals are developed to ensure that the treatment is administered with consistency to different clients by different therapists. Examples of empirically validated treatments are cognitive-behavioral manuals developed for depression, general anxiety, or obsessive compulsive disorder; parent training manuals for families with children who are diagnosed as being oppositional; manuals for behavioral marital therapy; manuals for applied relaxation for anxiety; manuals for brief dynamic therapy for depression; and manuals for systematic desensitization or exposure therapy for phobias and panic disorders.

However, there is growing criticism of using the criteria of empirical support (as currently defined) to be the primary guide in choosing interventions or conducting therapy more generally. This criticism focuses on the lack of attention that the kinds of studies or approaches used to establish empirical support give to aspects of therapy and the therapy relationship that are less easily operationalized quantitatively, the lack of attention to therapist variables, and the lack of attention to client variables and context. For example, Levant, & Hansen, (2008). point out that clinical judgment is multifactorial and difficult to measure quantitatively. The research of Dovidio and Gaertner (Dovidio, Gaertner, Kawakami, & Hodson, 2002; Gaertner & Dovidio, 2005) suggests that tasks between people from different racial backgrounds are affected by attitudes that they may be unaware of or even that are antithetical to their own self view (such as racial discrimination). This suggests that therapist and client variables are also likely to play a part in the process of therapy. Another example is the frequent criticism that samples for empirically supported therapy research typically consist solely or primarily of White European American, heterosexual middle-class patients.

Our knowledge about the strong effect of cultural, racial, and social class background on worldview, attitudes towards therapy, and ways of thinking, feeling, and behaving calls into question the validity of generalizing findings from such a narrow sample to all clients. Related to these critiques are epistemological critiques related to the validity or generalizability of findings that question the meaning of "rigorous scientific standards" which, in this research, is based in a positivist or post-positivist understanding which values objectivity, generalizability, and controlled circumstances (as opposed to individual or subjective experience, contextualized understandings, and an emphasis on how multiple interactions of people and context are impossible to control) (Addis & Walts, 2002; Chambless & Ollendick, 2001).

In other words, the applications of Empirically Supported Therapies (ESTs) likely require adaptation for differing populations, which could change the nature of the treatment and thus make the empirically supported claim potentially questionable. Alternatively, it is possible that the treatment is not actually the most effective for people who are different than those in the research study. Just as students who have different learning styles require a broad array of instruction techniques, clients have different capacities, orientations, and preferred domains. For example, clients with the same diagnosable disorder may benefit from different approaches given their motivation, responsiveness, and verbal capabilities.

Another consideration to take into account when evaluating research proposed to be most effective for particular disorders is the definition of the disorder itself. While the Diagnostic and Statistical Manual (DSM) describes the major symptoms of any disorder, there are still variations within each disorder with regard to etiology as well as manifestation of symptoms. In other words, diagnosis is still variable as evidenced in the studies where different clinicians are asked to "diagnose" the same patient and the researchers receive varied results (Corey, 2009).

In sum, the criticisms of Empirically Validated Therapy (EVT) or Empirically Supported Therapy (EST) approaches center on concerns of using a "cookbook" approach, where the perception of a particular symptom or diagnosis would then indicate the application of an EST, without a process of conceptualization. The step that is missing in the indiscriminate application of EST is that of conceptual formulation. Thus, there is increasingly a stronger advocacy for *evidence based practice* in guiding intervention choices (APA, 2006; Gallardo & McNeill, 2009), which would integrate research-based findings about the effectiveness of strategies and interventions with conceptualization of additional issues.

The Task Force on Evidence Based Practice from the American Psychological Association (APA, 2006) clarified the relation of EST and EBPP: "ESTs start with a treatment and ask whether it works for a certain disorder or problem under specified circumstances. EBPP starts with the patient and asks what research evidence (including relevant results from RCTs [randomly controlled trials]) will assist the psychologist in achieving the best outcome" (p. 273). In EBPP, the evidence comes from many sources including: direct quantitative research about the treatment of the presenting problem (empirically supported treatments); qualitative research about the treatment or the presenting problem; the application of research about similar types of clients (for example, research on cultural differences); theoretical and case example literature relevant to this type of client and presenting problem; common factors research; and your own experiences with other clients, similar presenting problems, and people in general. These multiple sources of evidence are integrated with the therapist's understanding of the unique client (starting with the patient), contributing to a conceptualization that utilizes all aspects of the therapist's knowledge and expertise while recognizing the uniqueness of the client. *A fourth major goal of this book is to help novice therapists develop skills for gathering and organizing information and knowledge and integrating this knowledge with self- and other-awareness into a case conceptualization that will contribute to evidence based practice.*

Conceptualization Is a Continuous Process

Conceptualization is a continuous process; the story is always changing with new information. We do not create an initial conceptualization of a client and have our actions dictated by that single "defining" statement. New information and circumstances come to light as we get to know a client, which forces us to continuously reassess and revise how we conceptualize him or her. As a continuous process, conceptualization is related to the ability to continually create, modify, and respond to contextual hypothetical thinking in clinical situations. It is also related to continuous evaluation of treatment, and the use of this evaluation as additional information to inform continuous conceptualization. *A fifth major goal of this book is to explore the process of iterative conceptualization.* How does one continuously gather data and modify conceptualization and interventions related to conceptualization?

INTRODUCING THE CASE OF NANCY

We want to introduce here a specific case that we will use (along with other clinical examples) throughout the following chapters to illustrate approaches to conceptualizing clients and therapy relationships. We will ask you to think about Nancy as if she is your client, asking you many questions to guide you in your conceptualization and treatment plan, rather than providing you with sets of possibilities.

The information below was obtained directly from Nancy during the first two sessions, which the therapist considered the initial intake. The initial intake is a time when the therapist gathers basic information (more on this in Chapter 7), while the therapist and the client are considering whether they would work well together, or if the client is best served by a referral to a different therapist or a different kind of treatment. We are assuming that you have already mastered the basic counseling skills for creating an initial alliance with the client and conducting initial interviews.

Referral Source

Nancy was referred to a community mental health center by her college academic advisor who had noticed a drop in the quality of her academic work and a change in her affect. This professor had originally suggested that Nancy go to the college counseling center, but Nancy did not feel comfortable discussing her personal issues with someone in the college. She was embarrassed that her advisor, who was the instructor of one of her current major courses, had called her in. Her mother obtained the name of a particular therapist at the community mental health agency that her professor had suggested. Nancy is covered by her family's health plan which enables her to have unlimited weekly sessions at this community health center.

Background

Nancy was a 20-year-old college junior at the time of her referral. She is a White European American of Irish American ethnicity on her mother's side and English American on her father's side. Nancy was raised in an affluent suburban area two towns away from the college she attends. She is the oldest of two—her younger sister has special needs that affect her academic and social functioning. Her father has a managerial position with a high tech company and her mother is considering a return to her freelance operations management consulting practice.

Nancy reports that her family is not religious although they observe Roman Catholic holidays with her mother's family. Nancy reports that her parents keep saying they were "self-made," that they had worked through high school and college and have no debt other than a mortgage, but they always seem to worry about money. Nancy reports that they live in a "nice" house and they rent a summer house yearly. To her mind, they spend money on themselves but not on her.

Nancy came from a privileged community where her values and attitudes were shaped by the materialism of her peers and by the larger community values of social and educational achievement. Because Nancy perceives that her family ranked in the lower socioeconomic quartile of this community, she was sensitive to "being different" and "not having as much as everyone else." She attended a private college where the same values of privilege prevailed.

Nancy's medical history is significant in that she underwent surgery at the age of eight to relieve pain from pressure that resulted from a treatment-resistant staph infection in her lymph node. At the time of her illness, her doctors and family did not believe her complaints about not feeling well. She states that they thought she was complaining about illness to avoid pressures at school or with her peer group because they thought that she had previously done this as well. As a consequence, Nancy reports that she is mistrustful of medical personnel. Nancy is currently healthy, with no problems noted by her physician. She works out everyday and is unhappy if she misses a session at the gym.

Nancy attends a private college, and her parents provide full college support financially, paying tuition, room, board, and other necessary expenses (for example, books). In spite of this, Nancy feels angry about the lack of things she has in comparison with her peers. In general, she perceives herself as deprived by her family (by her mother in particular) of financial comforts and believes that her sister is favored and treated better. Nancy has worked most summers and vacations in a retail store and uses her income to shop for luxury personal items as she thinks her parents should pay for everything else she needs or wants. She says that her mother is always complaining that they do not have enough money, but she does not believe her. She reports a conflicted relationship with everyone in her family but her father. Although Nancy cares for him and believes that he cares for her, she also criticizes him because he "thinks and does whatever my mother tells him to."

Nancy says that she cannot trust anyone and has never experienced an intimate friendship. She has not developed any "close" friends in college and feels alienated from her suitemates and classmates at college, although she does go out "drinking" with them on weekends and reports that she frequently drinks "too much." Socially, she relies on high school friends who return home from out of town colleges during vacations. Nancy describes that her socialization with these friends has also involved "partying" since ninth grade.

Nancy is involved in a sexual relationship (her first) with a taxi driver she met. She keeps this relationship a secret from everyone—family and peers—and never meets him where she might be seen by other people she knows. She says she needs him as he is the only person who cares about her and he is her "best friend." However, Nancy reported that she would be ashamed for her friends and family to learn about this relationship. She stated that she does not enjoy the sexual part of the relationship, but that she can "talk to him" and he is "supportive." In spite of this description, she actually seemed to know very little about this man or his life. She relied on him to be available whenever she wanted him, but would not give him her cell phone number or email. Although Nancy acknowledged that this would never be a long-lasting relationship, she felt that she needed to be wanted. She also admitted that she was avoiding meeting other men.

Presenting Concerns and Strengths

Nancy is concerned about her position in her family. Her mother is always criticizing her for being so "demanding," claims that the family's finances are strained, and that Nancy needs to learn to manage on her own. Nancy has worked during the summer and vacations and she has significant savings; she acknowledges that she spends a lot of money on shopping, an important activity for her. She keeps reiterating that her parents should pay for everything she thinks she needs.

While talking to a therapist is not something she is comfortable with, Nancy is concerned about her decreasing interest in her classes. At the same time, she is highly motivated to succeed as an economics major and she is applying for internships for the summer between junior and senior years. Thus, she needs to "buckle down" with her studies to be able to obtain a good recommendation from her advisor.

When asked what she perceives as her strengths, Nancy responds that she is responsible, a good student, and has self discipline. Because she has always been a good student, she feels that her parents pay more attention to her sister who has always had difficulties in school. When asked what she would like to have happen in therapy, Nancy is initially unable to respond.

Verbal and Nonverbal Behavior

Nancy was always well-groomed and appropriately attired. Physically, her eyes seemed sad and she often seemed to be tired. She appears to have good social skills in responding politely and interactively, but she rarely smiles, evidences no humor, and has little awareness of her effect on others. Her affect is bland, while her speech is measured and clear. Initially, her body language was stiff and reserved and she did not initiate conversation, but responded to direct questions and responsive listening. Over the first few sessions, Nancy's body stance became more relaxed and she initiated discussion more frequently. She is clearly intelligent and articulate; her verbal behavior reflects a concrete cognitive style. Her nonverbal behavior includes: twitching her foot, avoiding eye contact, and fidgeting with her sweater or scarf. These behaviors were most evident in initial sessions and in later sessions when Nancy was describing particularly difficult events. She is not able to identify feelings or talk about them in any depth. Her anger is focused on her mother and sibling, and she is quite adamant about her perceptions of their mistreatment. In the family session held during the first month, Nancy raged against her parents for not giving her what she wants materially. She angrily accused them of favoring her sister and demanded that they should buy her a car. She saw no reason why her mother should not run errands for her, make appointments, and so on.

Attitude towards Therapy

Nancy is concerned about how she will get to the community health center on a weekly basis as she refuses to take public transportation. She believes that "it's a waste of time" and she does not always have access to the family car. Her mother suggested she take a cab, but Nancy refuses to use her money in that way.

CONCEPTUALIZING NANCY

As discussed above, how you conceptualize Nancy and the interventions you would choose to help her are founded in your theoretical orientation, but tailored to your understanding of her specific issues, contexts, and presentation. Nancy can be conceptualized through any of the established theoretical orientations which fall into the general families of psychodynamic, cognitive-behavioral, existential-humanistic, and systems-ecological. In addition, constructivistic theories and liberation theories (feminist and multicultural) reflect philosophical perspectives that influence theoretical orientation and could be applied to Nancy. Appendix A on pages 232–249 provides a brief review of these orientations and examples of their application in conceptualization using the case of Nancy. We also provide some exercises to consider the relation of your own experiences and values to each of these orientations.

While we do not ascribe to any one theoretical perspective, we believe that each of the major models of psychotherapy has something of value to contribute to our understanding of clients and our efforts to help them. Most therapists do not ascribe to only one theoretical orientation, but instead develop a personalized integrative theoretical orientation. Different therapists are drawn to different theoretical orientations (or particular aspects of theoretical orientations) because of their own values, worldviews, and experiences, as well as because of the influence of the training they have received. The remaining chapters in this book focus on developing your understanding of your own theoretical orientation and its relation to conceptualizing clients in context, and applying this understanding to organize information for effective conceptualization.

SUMMARY

This book focuses on how you understand the process of change and how you apply this understanding to a particular client with sensitivity to the unique experiences and contexts of the client and to your unique relationship with each client. It also reflects an ecological approach to human development that celebrates the diversity of human experience. In this introductory chapter, we have focused on defining conceptualization; clarifying the difference between conceptualization and theoretical orientation; exploring conceptualization as a relational process; considering the relation of conceptualization to evidence based therapy, empirically supported therapy (EST), and empirically validated therapy (EVT); and asserting that conceptualization is a continuous process.

The goals for therapist trainees are: 1) to help them develop their own theoretical orientation; 2) to explore the process of conceptualization and influence of the trainee's theoretical orientation on the conceptualization process; 3) to identify the values, worldviews, and contexts that affect the therapist, the client, and the therapeutic relationship; 4) to help the novice therapist develop skills for gathering and organizing information and knowledge and to integrate this knowledge of the self and other awareness into case conceptualization for evidence based practice; and 5) to explore the process of iterative conceptualization.

In this chapter, we also introduced the case of Nancy, which will be referred to throughout this text book to illustrate the issues that are discussed.

REFERENCES

Addis, M.E. & Waltz, J. (2002). Implicit and untested assumptions about the role of psychotherapy treatment manuals in evidence-based mental health practice. *Clinical Psychology: Science and Practice* 9(4), 421-424.

Antony, M.M. & Barlow, D.H. (Eds.). (2010). *Handbook of assessment and treatment planning for Psychological disorders*. New York: Guildford.

APA presidential Task Force on Evidence-Based Practice (2006). Evidence-Based practice in Psychology. *American Psychologist*, 61, 271–285.

Berman, P.S. (2010). *Case conceptualization and treatment planning: Exercises for integrating theory with clinical practice*. Thousand Oaks, CA: Sage Publications, Inc.

Castonguay, L.G. & Beutler, L.E. (2006). Principles of therapeutic change: A task force on participants, relationships, and techniques factors. *Journal of Clinical Psychology* 62(6), 631-638.

Chambless, D.L. & Ollendick, T.H. (2001). Empirically supported psychological interventions: Controversies and evidence. *Annual Review of Psychology* 52(1), 685-716.

Corey, G. (2009). *Theory and practice of counseling and psychotherapy*. Brooks/Cole Pub. Co.: Belmont, CA.

Dovidio, J.F., Gaertner, S.L., Kawakami, K., & Hodson, G. (2002). Why can't we all just get along? Interpersonal biases and interracial distrust. *Cultural Diversity and Mental Health* 8, 88-102.

Gaertner, S.L. & Dovidio, J.F. (2005). Understanding and addressing contemporary racism: From aversive racism to the common ingroup identity model. *Journal of Social Issues*, 61, 615-639.

Gallardo, M., & McNeill, B. (Eds.). (2009). *Intersections of multiple identities: A casebook of evidence-based practices with diverse populations*. New York, NY: Routledge.

Germain, C.B. (1979). *Social work practice: People and environments, an ecological perspective*. New York, NY: Columbia University Press.

Hersen, M., & Porzelius, L.K. (2002). *Diagnosis, conceptualization, and treatment planning for adults: A step-by-step guide*. Mahwah, NJ: Lawrence Erlbaum.

Levant, R.F.; Hasan, N.T. (2008). Evidence-based practice in psychology. *Professional Psychology: Research and Practice*, 39, 658–662.

Norcross, J.C. (2002). Empirically supported therapy relationships. In J.C. Norcross (Ed.), *Psychotherapy Relationships that Work: Therapist Contributions and Responsiveness to Patients* (pp. 3–16). New York: Oxford University Press.

Sperry, L. (2005). Case conceptualization: A strategy for incorporating individual, couple, and family dynamics in the treatment process. *American Journal of Family Therapy* 33, 353-364.

Stevens, M.J. & Morris, S.J. (1995). A format for case conceptualization. *Counselor Education and Supervision* 35(1), 82-94.

Wampold, B.E. (2001). *The great psychotherapy debate: Models, methods, and findings*. Mahwah, NJ: Lawrence Erlbaum.

2

Exploring Influences of Personal Worldviews

Our goal in this book is for you to be able to formulate your own case conceptualization, grounded in your own theoretical orientation, including awareness of what you draw from established theories, how and why you make those choices, and how your understanding affects treatment decisions. In this chapter, we explore the influences of personal worldviews on theoretical orientation and effective case conceptualization. We discuss the importance of theoretical orientation, examine your initial understanding of how to create psychological change, encourage reflection upon your own values and your understanding of health and pathology that may affect the development of your theoretical orientation, and introduce dimensions of understanding that can help you develop an integrative theoretical orientation and apply this to an effective case conceptualization.

WHY IS THEORETICAL ORIENTATION IMPORTANT?

Research suggests that therapists' specific theoretical orientation does not seem to account much for the effectiveness of psychotherapy, although there is much evidence that therapy is effective in facilitating positive change (Wampold, 2001). Therapists who vary in theoretical orientation may, therefore, do very similar things, while understanding or explaining the rationales for their actions quite differently. If this is so, a novice therapist might ask: Why do I need to develop or understand my theoretical orientation? We believe that people benefit from having consistency in their identities and reflexive understandings about the goals and motivations of their actions, particularly when they are engaged in purposeful activity. Research has also shown that therapists believe that theoretical orientation strongly influences their actual practice, and that clinical experience and personal values, as well as professional training, affects the development of theoretical orientation (Norcross & Prochaska, 1983). Thus, while the content of various theoretical orientations may differ considerably and the effect of a specific claimed orientation may not be evident, we believe that the process of considering, applying, and modifying our theoretical orientations in our work with clients contributes to being mindful and intentional, which contributes to our effectiveness as therapists.

However, it can be challenging for novice therapists to decide what theoretical orientation makes the most sense to them. Furthermore, many therapists do not adhere to a single theoretical orientation. They seek instead to develop an integrative approach, which reflects their personal relationship to theory about change. Similarly, few therapist trainees adhere solely, or even primarily, to a single theory but prefer to develop an integrative approach.

EXERCISE 2.1 ■ Consider the established theoretical orientations reviewed in Appendix A. Assign a different theoretical school to each member of a six-person small group: psychodynamic, cognitive-behavioral, existential-humanistic, systems-ecological, constructivistic, and liberation. Each person should advocate for the position of his or her theoretical school. Can you agree on a single best theoretical orientation? Why or why not?

EXERCISE 2.2 ■ Identify two theories that you like and give at least two reasons why you chose each of these theories. Then, identify two theories that you do not like, giving at least two reasons why you do not like each of these. Compare and contrast your responses with two of your peers.

Consider your answers to Exercise 2.2 above: Are you describing the parts of each theory that you like, or are you able to describe *why* you like different theories in relation to an understanding of how change develops (for example, "I like the emphasis on insight in psychodynamic theory because…")? Our experience is that novice therapists frequently focus more on the content of the theories of therapy and less on how these theories interact with their own experiences and worldviews, that is, on why they adopt, adapt, or reject established theories in their own practice. Trainees can frequently say: "I like X about this theory, but I don't like Y; and I like Z about this other theory." However, when asked *why* they have these preferences, trainees are frequently stumped; they have no framework upon which to determine an understanding of their own understanding. They struggle with trying to fully integrate different theoretical or technical aspects of theories because it is difficult to articulate the beliefs about change or people that attract them to different aspects of theoretical orientations. Even trainees who ultimately choose a single established theoretical orientation are not always aware of why they are choosing that orientation. But an awareness of this interaction is crucial, so that you may be able to effectively apply and modify your theoretical orientation to the conceptualization of a specific client.

UNDERSTANDING YOUR BELIEFS ABOUT CHANGE

The purpose of psychotherapy is, of course, to produce beneficial changes in the lives of clients. While this is the goal of all psychotherapy, therapists differ in their understanding of change and the best ways to bring about change. Theoretical orientation reflects the answer to the basic question: What is *your* theory of psychological development and change? This question incorporates many other questions such as: What is change? Is change natural? Is it developmental? Is it inevitable? Is it behavioral? Cognitive? Emotional? Is it all of these? Does one kind of change precede another in a sequential fashion? Do developmental change and intentional change occur in the same ways? Does first-order change inevitably relate to second-order or "deep" change, or does something else have to happen?

Making choices about theoretical orientation means being aware of where your beliefs come from, so you can question and consider whether these beliefs can or should be applied to all clients or to only some clients, and in what circumstances. For novice therapists, evaluations of what is most or least likely to work in creating change in a client's life is likely to be related to personal worldviews and experiences, as they have little experience with actual clients upon which to base their interventions and understandings. Intially, your theory of how therapy might create change is likely to be related to your experiences of change in your own life. Therefore, we turn now to considering the question: How do I change?

EXERCISE 2.3 ■ Think about the person you were five years ago. How are you different now than you were then? Consider some of the following:

- your personal characteristics
- the relationships you have with family members
- what you look for in friends and intimate relationships
- what kinds of things create stress in your life
- what kinds of things you do for enjoyment or relaxation

- your educational aspirations and career goals
- the things you would like to change about yourself or your life.

Make a list of at least three things that are different in your life between then and now. For each thing in your list, describe some of the influences that contributed to the change. These influences may include people, new perspectives or ways of thinking, or new activities in your life. Consider also whether the change was intentional or unintentional. Compare your list with those of at least three other students in a small group. What are the similarities and differences in areas of change? In influences upon change?

EXERCISE 2.4 ▪ Consider a time in your life when you intentionally and actively changed something about yourself or your life. All of us have gone through change, such as changing study habits, breaking up with an intimate partner, switching schools or majors, losing weight, stopping smoking. With a class partner, describe the issue and how you went about making this change. Your partner should then imagine being in a similar situation and describe how she or he would have approached changing the specific issue.

EXERCISE 2.5 ▪ Imagine yourself in one of the following scenarios:

1. You have an old friend with whom you used to spend a lot of time. However, over the last year or so, it has felt like a lot less fun to hang out with her. It is hard to figure out what to do when you are together and you both seem to share less of what is going on in your lives with each other. You feel that she is frequently critical of you and argues against whatever you say. You frequently feel tired and unbalanced after seeing her.

2. Schoolwork used to be fairly easy, but recently it has been much harder. The material seems much more difficult than it ever was. Even though your grades are slipping, you find yourself procrastinating whenever it is time to study or write a paper. You are worried that you are going to let your family and friends down, as they have always been proud of your educational accomplishments.

3. You would like to lose some weight and you know that you feel better when you exercise regularly. In addition, at your last physical check up two months ago, your doctor said that you had high blood pressure and high levels of cholesterol. She recommended some beneficial changes in your diet and increased exercise, but somehow you are not following them.

Write a paragraph expanding upon one of these scenarios and your experience. Then, describe your thoughts on how to address the problem. If you imagine yourself in the scenario, what are some of the causes of the difficulties of changing? How would you approach the problem? Compare your description with those of at least three of your peers. How do you differ from each other in relation to imagining yourself in the scenario? What aspects did you expand upon? How are your approaches to change similar to or different from each other?

EXERCISE 2.6 ▪ Now, after considering your answers to the exercises above, write down five ideas or thoughts about what change is and how it happens. These ideas must reflect your own ideas and not be simple recitations of what you have been formally taught. When you have completed this list, discuss your ideas or thoughts with at least three other students in a small group. As a group, compose a master list with a column of shared ideas and a column of individual or unique ideas. Remember that there are no right or wrong ideas! Some questions to consider are:

1. Is change sudden or spontaneous? Planned? Persistent?
2. Is change temporary? Permanent?
3. Is change consistent and ongoing? Static?
4. What roles do motivation, learning style, or attention play?
5. What kinds of changes are easier than others?
6. What or who has been able to help you change and how?

As you process these exercises with your peers, pay careful attention to the cultural diversity within your group, to influences that may relate to gender, ethnicity, race, religion, class, and sexual orientation.

INFLUENCES ON YOUR IDEAS ABOUT CHANGE

There are multiple influences on your understanding of people and change. Some of these are related to the unique constellation of your individual and family experiences, such as the particular school you attended, teachers and friends you had, and the hobbies you pursued. Other major influences are less individual, and even what we view as an individual influence is frequently related to shared experiences and understandings.

Cultural Influences on Understanding Clients: Beliefs about People and Worldview

Every therapist has his or her own understanding or worldview about human nature, about the development of health and psychopathology, and about the meaning and process of change both generally and specifically through psychotherapy. This worldview is like a preferred lens through which the therapist develops understanding of the client, his or her issues, and the contexts in which he or she exists. This understanding enables the therapist to formulate the best treatment approach. Thus, our awareness of the existence and nature of this lens is necessary to understand how we can be most effective. In addition to unique personal experiences, one important influence on the worldviews of therapists and clients is culture. Some of the variables that differentiate cultures and individuals from different cultures, and that affect the therapeutic relationship and understanding of change include the nature of human beings and their influence on environment, communication styles, and the nature of self and relationships. (See Appendix B for more detail of these variables.)

Most therapist trainees receive specific training about cultural differences and the ways that ethnic and racialized minority groups may differ from the White European American culture that is dominant in the United States. While early training in therapy frequently includes some emphasis on personal and cultural self-awareness, trainees do not always see how these issues shape their theoretical orientation and the related case conceptualization. Examples might include: if your culture (and you) believe that people have an inherent tendency towards self-actualization and emphasize the value of emotional awareness, you may be more attracted to existential-humanistic theories; if you value hierarchical relationships and expert advice, you may be more attracted to more structured and directed therapies rather than, for example, client-centered therapy.

EXERCISE 2.7 ▪ What are some of the cultural norms and beliefs from your culture that might be particularly relevant to influencing your beliefs about how change happens, about what makes problems develop, or about how and why therapy is helpful?

Power, Privilege, and Inequity

Some of the variables that relate to culture and other ecological systems also relate to systemic inequities such as race, gender, sexual orientation, and social class. These inequities influence individuals' experiences with power and privilege or with (relative) powerlessness and oppression and also affect our understanding of mental health and mental illness, clients' presenting problems, and conceptualizations affecting interventions. By *power*, we mean the ability to influence circumstances (and people!) in a desired direction. By *privilege*, we mean having preferred status in a social system of hierarchy that benefits some, but not others, in ways that are not connected to effort or ability. People in privileged spaces have more power because they not only have the power that they earn/create, but also the power that is given to them (and not to others) because of their status.

Systems of power and privilege that are of particular importance in the US context that affect clients and therapists include race, gender, social class, nationality and ethnic origin, sexual orientation, and religion. What sets these apart as systems (rather than individual differences) is that there are social understandings and histories that create inequities in power based on one's status in relation to these variables, and not on one's own efforts.

Most therapist trainees have also received some specific training related to issues of power and privilege as part of their training in multicultural competence. However, while this training almost always encourages trainees to recognize inequities in power and privilege and the effects of these upon clients' experiences (particularly those of minority clients), many have not fully considered how their own experiences of power and privilege in relation to multiple intersecting statuses have affected their understanding of how people change, what constraints exist in relation to change, or how they interact with clients who have different experiences with power and privilege. The majority of trainees (and people) have areas or statuses in which they are privileged and areas where they have experienced relative oppression. The areas in which you have systemic privilege are most likely to be the areas that you have not fully considered your own experiences and their effects. It is also important to remember that historical legacies still affect current experiences and that these experiences affect interactions in psychotherapy. (Vasquez & McGraw, 2005).

EXERCISE 2.8 ▪ What kinds of daily benefits and experiences do you have or not have based on your status within systems related to power and privilege? How does this affect your view of yourself, of people, of opportunity? How has it affected your individual and family psychological dynamics or health? How might it affect the choices you make about theoretical orientation or the ways in which you might conceptualize the psychological development and change process for someone with similar or different experiences?

Although it is not the purpose of this book to fully explore the influence of culture and other ecological systemic variables on the therapist and the therapy, it is vitally important that therapists understand their own cultural backgrounds, as well as their own experiences with power, privilege, oppression, and (relative) powerlessness because these do actively shape one's theoretical orientation and approach to therapy. For example, people who are mostly in privileged statuses may be more likely to think that their experiences are related to their own efforts and characteristics, and that anyone can change almost anything psychological if they really want to. In contrast, people who have experienced oppression may be more familiar with the environmental or interpersonal constraints that such systems place upon individuals who are trying to change. (See, for example, the discussion of locus of control versus locus of responsibility in Sue & Sue, 2007). If you have not yet fully explored your own cultural background and experiences of power, privilege, and relative oppression, we strongly encourage you to do so. A review of related material is included in Appendix B.

EXERCISE 2.9 ▪ Review your answers in Exercises 2.3 to 2.6 about your process of change. How might your cultural values, your beliefs about people, and your experiences of power and privilege affect how you have changed? How might your experiences be different if you had significantly fewer resources (much less money, fewer friends, lack of family support)? Or if you had numerous experiences of oppression due to racism, sexism, homophobia, or some other systemic inequity?

Power and Privilege Related to the Role of the Therapist

Power and privilege relate not only to social statuses as described above, but also to the different roles that people occupy. There is a difference in power and privilege between teachers and students, supervisors and trainees, therapists and clients. How you have personally experienced and how aware you are of your own power and privilege in relation to your social statuses may also affect your understanding and willingness to address issues of power and privilege in relation to your role. People in privileged roles—therapists, teachers, parents, and clergy—have more power than clients not only because of their knowledge and expertise, but also because of their status. It is critical that we acknowledge the power and privilege of our role and consider the interaction of our role as therapists with our status on social variables so that we do not abuse our role power. By *abuse of power*, we are referring to the imposition of our values, attitudes, and beliefs upon others of lower power and privilege based on the belief that what we think, value, and believe is normative or of greater value.

EXERCISE 2.10 ▪ In small groups, describe an experience when you have felt relatively powerless in relation to your role—for example, as a student, an employee, a customer, or a family member. How did power and privilege related to your social statuses interact with your role in this experience? What is it like when you are in a group where you feel different as opposed to where you feel similar to others? The purpose of this discussion is not only to get in touch with your own feelings of comfort and discomfort (and the resulting thoughts and behaviors), but also to learn about others' experiences.

The way that you understand the power of the therapist will also interact with your understanding of change and how best to facilitate it. A question that relates to your understanding of the therapist's power, as well as to your view about the nature of people generally, is whether the knowledge necessary to create positive change is primarily located in the client or in the therapist. If you believe that people inherently have a tendency towards self-actualizing and see the therapist's role primarily as a facilitator of this tendency, then you are more likely to see the knowledge necessary for change as residing within the client. If, on the other hand, you believe that people do not have such a tendency and that the therapist's role is to share expert knowledge that the client does not have in order to facilitate change, then you are more likely to see the knowledge necessary for change as residing within the therapist. Most therapists are not on either end of this continuum, but fall somewhere in the middle. People who are more privileged are more likely to have had their worldviews and values validated and appreciated by those in power (such as teachers and therapists) or by the institutions with which they interact (such as schools, media, government). Therapists like this may be more comfortable with asserting their expertise or more cautious about doing so because they are particularly aware of the danger of imposition.

The question of the location of knowledge also reflects your views on epistemology, which addresses the questions: What is knowledge? What is valid knowledge (truth)? How do we know that we know?

EXERCISE 2.11 ▪ How do you know what you know? What makes you confident that something that you know is true? Consider your answers to Exercise 2.7. How do you know that your views reflect how people really are? How do you account for views that differ from your own (are these mistaken, lacking information, equally valid)? Think about a recent time when you learned something that challenged what you had previously thought to be true. What was this process like? How did you go about evaluating the different views or information? What convinced you to change your understanding?

Some people are strongly influenced by their own personal experiences or the personal experiences of others. Some are more strongly convinced by statistics and scientific studies or expert opinion. (See Sue & Sue, 2007 for a related overview of the basis of credibility within therapy.) Most of us are influenced by some mix of these influences. Your epistemological philosophy will affect how you learn about therapy and how you interact with your client. You will have knowledge and expertise about people and psychological functioning that your client does not have. Your client will have knowledge about him or herself and likely about people similar to him or her (such as others from the same culture or community) that you do not have. How does this knowledge and expertise come together to contribute to positive change in the life of your client? Your views on this will affect your position on many of the dimensions that we will address in this and future chapters. It will also affect your affinity for aspects of traditional theoretical orientations, as these, too, have epistemological philosophies behind them.

Your epistemological philosophy will also have a strong influence on how you understand the meaning of health and pathology, and what makes a behavioral, thought, or emotional pattern problematic and, therefore, something that should be addressed in therapy. We now turn to explore this more fully.

BELIEFS ABOUT HEALTH AND PATHOLOGY: THE SOCIAL CONTEXT OF PSYCHOTHERAPY

Historically, psychological theories and research related to mental health/illness and to psychotherapy have been developed in European or European American contexts. The modal and social standard criteria were based on European or European American populations where men, rather than women, were viewed as the epitome of healthy human beings. In the last few decades, there has been increasing attention and criticism about applying standards of normality and health created by and for White European or European American men to women, people of color, and ethnic minorities in the United States, as well as to individuals and societies across the world.

While cultural relativism is an accepted element in mainstream psychotherapy today, one must also consider contexts and intentions when deciding what is healthy or pathological. For example, a high school physical education teacher notices what looks like multiple severe round bruises on John's back. The teacher is concerned about possible child abuse and notifies the school psychologist. As John refuses to talk about the bruises, the psychologist notifies child social services for the state. When investigated, it becomes clear that the "bruises" are the result of the traditional Chinese healing practice of cupping, where suction is created by the application of glass or plastic cups to draw the toxic elements from the qi in the body. While leaving marks similar to large "hickeys," most people find the actual cupping experience to be slightly uncomfortable, but not painful (which is similar to many Western healing techniques which are rarely lots of fun to encounter).

Alternatively, the physical education teacher in Jan's school also noticed her bruises. Child social services interviewed her family, who explained that in their culture, if a girl was seen with a boy outside of the family purveyance, it was considered a disgrace and physical punishment was indicated. In these cases, the bruises were the result of behavior that was perceived as acceptable and right within the cultural context. However, the acceptability of the behaviors also needs to be evaluated within the current context and ethical and legal evaluations of health, pathology, and acceptability. For John, the intention of the behavior was healing and he understood it as such. In this case, the social worker judged the bruises to be not the result of abuse, but did speak to the parents about the need to possibly alert and educate school personnel in the future. In Jan's case, the intention was not healing and the bruises had a very different meaning. A family intervention plan was developed to educate Jan and her family about the norms of the society in which they now reside. While in their home culture, the behavior and intention were not considered "pathological," however in the United States, it is perceived as both pathological and illegal.

Historically, established theories of mental disorders did not have the scientific knowledge we have today about the biological conditions (such as increasingly new understanding of the role and function of the brain) pertaining to mental health. The established theories were based more on the psychological conditions experienced cognitively and emotionally in our conscious and unconscious life. Today, for example, we know about the different impact of hormones upon the male and female brains, which may relate to mood disorders. Furthermore, in recent years we have come to appreciate the contexts of the larger social systems affecting our health and creating the cultural norms, values, and morals affecting such matters as responsibility and personal agency. While there is still controversy about nature versus nurture, new findings enable us to continuously reassess our assumptions, recognize complex interactions, and modify our conceptualizations about all aspects of human behavior.

Most behaviors exist on a continuum in terms of what is considered "normal" or "pathological." A perennial question for therapists in conceptualizing individuals is how to draw the line between acceptable variability and pathology. For example, the vast majority of people experience some anxiety in some situations. In fact, a certain level of anxiety is known to be beneficial, as it focuses our attention and alerts us to be prepared. To illustrate the continuum, let's imagine a simplified example of some students who are about to take an exam the following day. All of these students have attended every class and read the material as it was assigned.

- Larry is not at all nervous or worried about his performance on the exam; he has attended all the classes and he spends the evening watching television and cruising YouTube videos.
- Joe is a little nervous and spends the evening reviewing the book and his class notes, plays a few video games to relax, and then goes to bed around midnight and sleeps soundly.
- Adam is fairly nervous; he stays up late studying, making new outlines of concepts and materials, eventually going to sleep around 3:00 AM and dreaming about not having the right pen to take the exam.
- Mike is very anxious; he pulls an "all nighter" and develops detailed notes of every concept, and spends hours spell checking, reorganizing, and re-wording his notes. His efforts are frequently interrupted by his need to go to the bathroom to throw up.

It is likely, assuming relatively equal abilities and past experience with the topic, that Joe and Adam will perform better than Larry and Mike. The question for the therapist is: at what point is the level of anxiety no longer acceptable and how do you determine this? Related to this decision is a question of how one understands the problem in relation to the context.

Criteria for Evaluating Health and Pathology

Four possible criteria for distinguishing health from pathology are 1) judging what is pathological according to modal or average experiences; 2) judging what is pathological according to accepted social standards; 3) judging what is pathological according to authorities; and 4) judging what is pathological based on what is considered dysfunctional for individuals generally. The extent to which you rely on each of these criteria (or others) is related to your epistemological philosophy, as well as to your own experiences and beliefs about change, cultural worldview, experiences with power and privilege, and understanding of the role of the therapist.

Judging What Is Pathological According to Modal or Average Experiences This involves comparing the behavior or experience of one person with the experience of a group of people. For example, if the vast majority of students throw up when preparing for an exam, we would think differently about Mike's behavior. Many psychological tests and diagnostic criteria are related, at least partly, to normative criteria for health—pathology is judged by behavior or experience being unusual.

An issue in utilizing this criterion is determining the appropriate reference or comparison group. Many students do not study as much as Adam. Should we compare Adam's behavior to all students or to some subgroup of students? Adam might be a scholarship student from a refugee family who is spending their life savings and taking out large loans to enable him to pay for college expenses, rather than contribute to the financial situation of his family. His performance may relate to whether he will continue to receive his scholarship and be able to stay in school and whether he will be able to support his family in the future. On the other hand, Larry may be independently wealthy and set to take an executive role in his father's company regardless of how well he does in college. Should we compare Adam's behavior to that of all students or just students from similar backgrounds? What groups should be used for social comparisons? Is the majority necessarily healthy because they are the majority? These questions are particularly important when considering that different cultures vary with regard to what is modally normative in values, communication, interpersonal interactions, and many other variables.

Judging What Is Pathological According to Accepted Social Standards Judging what is pathological according to accepted social standards may sometimes be influenced by modal or average experiences, but it is not exactly the same. For example, it might be unusual for someone to be a millionaire, but we do not consider being a millionaire pathological! Judging behavior on the basis of social standards means comparing individuals to a socially accepted idea of what is within the range of normal or positive behavior. Even if most students do not study as much as Adam, would we judge his behavior as abnormal or pathological? An issue in utilizing this criterion is that our judgment of acceptable behavior may vary with the context, just as our selection of a comparison group may be affected by context. Is Adam's behavior healthier because of his particular context? Do we apply the same social judgment of health to all people in all situations? Who determines the social standards? How? An example of the complexities in the method of judgment is reactions to intense trauma, such as the Jewish holocaust during World War II or the genocide in Cambodia under the Khmer Rouge. Are post-traumatic symptoms in survivors of these traumas pathological or a "normal" response to abnormal conditions?

Judging What Is Pathological According to Authorities Judging what is pathological according to authorities involves experts setting criteria for what is normal versus what is pathological. These authorities may base their opinions on empirical research, social standards, and modal experiences; on their own experiences; or on the basis of something else. We trust them to tell us what is acceptable and what is not. An issue in utilizing this criterion is that experts, like all human beings, are influenced by their own cultures, values, education, and experiences. Thus, the judgment of experts is not necessarily objective or unbiased. An expert who was similar to Mike in his anxiety about exams may view Mike's behavior as unusual, but not pathological. Who should have the power to decide whether a behavior is pathological or normal? Who is considered an expert and what qualifications should an expert have (or not have)?

Judging What Is Pathological Based on What Is Dysfunctional for Individuals Generally Judging what is pathological on the basis of what is dysfunctional for individuals involves considering the effect of the behavior on their lives and experiences. Behavior is pathological if it has a significant detrimental effect on individuals. One issue in utilizing this criterion is that what is dysfunctional for one individual may not be for another in a specific context, but judgment is usually created for individuals more generally. Thus, while this criterion has a greater sensitivity to context, there is still frequently a general, less contextualized assumption about what is functional. Another issue in utilizing this criterion is whether it is the individual or someone else (such as the therapist or society) who judges the dysfunctional effect. Mike may really be distressed by regularly throwing up while studying for an exam. But it is possible that this action does not distress him at all. If it does not distress him, is it dysfunctional? What if this action is causing physical problems such as decay of tooth enamel?

EXERCISE 2.12 ▪ In your own experience, what makes you decide that something you are doing, feeling, or thinking is an issue that you would like to address? What makes you decide that something is a "problem"? At what point would you consider that problem to be "unhealthy"? And at what point would you consider it a "pathology"? How does thinking about this experience as an issue you desire to address, as a problem, as unhealthy, or as a pathology, affect how you feel about it? How you feel about yourself? What actions and effort you might engage to address it?

EXERCISE 2.13 ■ In small groups, discuss the following scenarios. Decide where would you draw the line between "normal" and "not normal" and what criteria you are using:

a) Lili believes she has the right to express her feelings and screams at her 7-year-old son when he does not behave in the ways she wants him to. When her husband tries to intervene, she tells him that he should be yelling at their son instead of being so "passive." Now, when her son does something "wrong," he hides in his closet in fear of his mother's anger.

b) Paul, age 38, avoids confrontation and does not like to "make waves." When his 9-year-old son pulls his wife's hair, bites, and scratches her, he walks away.

c) Marcy, age 17, is enraged at her boyfriend's "abandonment," follows him around at school and continuously sends texts to him. Marcy's parents are divorced (her father disappeared from her life when she was 9) and her alcoholic mother appears not to be interested in anyone but the men she brings home.

d) Jill, age 52, is a working class single parent whose daughter has been awarded a scholarship to a prestigious college. Jill has started having "headaches" and other physical symptoms and tells her daughter that she knows she has some kind of cancer that the doctors have not yet found and that she is going to need her daughter to stay home and nurse her.

e) Marla, age 49, has been considering divorce for many years. When her husband announces he wants a divorce (before she had made her decision), she is outraged and cries incessantly every time she sees him. She claims she cannot sleep and that she has to contain her desire to hit him every time she sees him.

f) Len, age 58, lost his wife of 28 years due to pancreatic cancer a year ago. He is now having difficulty sleeping, eating, and focusing on his work.

EXERCISE 2.14 ■ Refer to the case of Nancy and identify three issues that you see as problems. On what basis do you make that judgment? As you consider these scenarios, how aware are you of your thoughts and feelings? How do you want to intervene in these cases? What contextual cues influence your thinking?

Influences on Your Conceptualization of Health and Pathology

Diagnostic and Statistical Manual (DSM) Therapists' understandings of health and pathology are not just individual decisions, but are strongly influenced by the norms in our field, and particularly by the Diagnostic and Statistical Manual (DSM, 2000). The DSM, part of the larger ICD medical codes, attempts to integrate the four approaches determining pathology. The diagnostic criteria are determined and codified by experts through considering what is unusual, what is considered socially unacceptable, and what is considered harmful. Each diagnosis has a criterion of significantly affecting the functioning of an individual (although the individual may not be fully aware of the detrimental effect). The primary goal of the DSM is to inform treatment through offering a tool for empirical assessment and treatment planning. Even so, there are many diagnoses, both historically and currently, that have significant controversy about them, and the DSM has been criticized for not attending enough to cultural and social influences on judgments or to the possible functions of behavior for clients in context.

Cultural Influences on Understanding Health and Pathology Homosexuality is a classic example of the way that the determination of pathology is culturally biased, and of how the DSM may contribute to pathologizing differences in behavior. Until 1973, homosexuality was a psychopathological diagnosis in the DSM. This reflected all of the criteria for judging pathology described above: most people do not identify themselves as homosexual; it was considered socially unacceptable; experts agreed that it was pathological; and it was seen as having a detrimental effect upon the life and happiness of individuals. The difficulty is that *all* of these judgments were influenced by social attitudes and values. We cannot know, for example, how many people would embrace same sex intimacy feelings if we did not live in a society that privileges heterosexuality and denigrates being gay, lesbian, or bisexual. Similarly, current research indicates that negative effects of being gay, lesbian, or bisexual relate to the social attitudes and homophobia rather than to inherent homosexual feelings, actions, or identifications (APA, 2011).

Cross cultural research also highlights the necessity of examining the ways in which our ideas of health and pathology are culturally determined. For example, in US society, we consider it healthy to be independent and autonomous, to make personal decisions based on our own choices and preferences, and to be personally assertive and achieving. In Japanese society, they consider it healthy to put the group interests before those of the individual, to be deferential to others, to be humble and self-effacing (Markus & Kitayama, 1991). Therefore, behaviors that may be functional in one cultural context may be dysfunctional in another. Individuals with cultural influences from Japanese society who live in US society (such as Japanese immigrants or Japanese Americans) may be judged as passive and overly dependent on family or others and may have difficulty with career advancement because of these judgments (a detrimental effect on life functioning). But people in Japan, Japanese immigrants, or Japanese Americans may also judge some actions of European Americans as rude, selfish, and aggressive (Nagata, 1993).

Controversies about the DSM and the specific diagnoses described within it continue with each version. Critics question the basis of judgments of pathology, raising questions similar to those just discussed. For example, as the fifth edition is being prepared, there is controversy about whether binge eating disorder should be included separate from other eating disorders, about the inclusion of gender dysphoric disorder, whether grief that lasts longer than six months is pathological, and about other diagnoses currently included or being considered for addition. These dialogues and criticisms are vitally important as they maintain our awareness that our judgments are shaped by our own contexts, cultures, worldview, and biases. This awareness can contribute to our positive use of diagnosis in our efforts to help clients (see Chapter 7).

Individual Functional Influences on Conceptualizing Health and Pathology We must also consider the *specific* and unique personal, interpersonal, and systemic contexts of clients and the possible functions that their behaviors, feelings, and thoughts may have for them and for others. Consider the following case:

> Alys, age 14, was admitted to the child psychiatry inpatient unit because she had been raging against her mother and had tried to jump out of a moving car when her mother was driving her to school. Clearly angry, Alys complained that her mother would not let her see her father (the parents were divorced) and that her mother was continuously trying to poison Alys's mind about her father. Alys's father was out of work and unable to afford to go to court to obtain visitation rights. The inpatient staff diagnosed Alys as having a conduct disorder and recommended a family session with both parents. During this family session, the treatment team was amazed at the level of rage and acrimony between the parents, particularly expressed by the mother. Observing this parental behavior raised questions about the meaning of Alys's behavior and the appropriateness of the diagnosis, reflected in the moment when the family consultant turned to Alys and said "I can understand why you wanted to get free."

Exploring the family context led to a different view of Alys's behavior by considering the function that it may serve for her and for her family. Seeing Alys's behavior in context enabled the team to arrange a supportive discharge plan where the family system became the client rather than the child. While assigning a diagnosis may sometimes require decontextualizing behavior and simply considering whether it is present or absent and affecting functioning or not, a full conceptualization must go significantly beyond that and even differential diagnosis is usually affected by the meaning and intent of behavior.

In sum, we are encouraging you to carefully consider the basis of your determination of health and pathology, attend to your own values and bias, recognize the culturally situated biases of the field, and consider the particular and unique contexts of the client. Like the DSM, most therapists will integrate the four criteria described above in their approach to evaluating pathology and mental health. What is most important, however, is to examine our personal tendencies for judgment and question the bases of our judgments.

INTRODUCING DIMENSIONS OF CHANGE

The discussion above is focused on illuminating how your understanding of change and/or health and pathology have been shaped by your life experiences, particularly your culture(s) and experiences with power and privilege. You may be aware of some of your values and preferences; others may be inferred by the choices you make and by your behaviors. These concepts influence how you understand the process of change in psychotherapy and the choices you make about how to facilitate this change. What you actually do, your "theories of use," may be more reliable indices of your concepts than your "espoused" endorsement of a therapeutic orientation. Thus, an additional way to become aware of your worldview (and its relation to understanding change) is to observe yourself and actively consider both alternatives to how you behave and why you make the choices you do.

While your worldview affects your understanding of change, it need not be deterministic. A therapist's understanding of change is not applied indiscriminately or rigidly. It is similar to a personality style: most people have a preferred personality style, which is what is measured in personality inventories. Healthy people, however, are not locked into this one style; they change their approach depending upon the demands of different situations. Similarly, good therapists recognize that their preferred lens (theoretical orientation guiding conceptualization) is not the only way to view a client or approach change. A therapist's preferred worldview or approach to therapeutic change can be significantly modified or even discarded according to the needs of a particular client. Usually, however, the therapist starts with and continues to use his or her preferred lens unless there is a clear reason to shift. In addition, as you gain experience, you continually modify your observing and self-reflecting on the interaction of self, client, and therapeutic process. You also become more aware of your beliefs about people and change and how these beliefs shape your approach to being a therapist.

It can be challenging for novice therapists to make connections between their values, preferences, relational styles, and their theoretical orientation as therapists. The connections between personal preferences/choices and theoretical aspects can be difficult to understand because each theory seems to be a complete package. Therefore, it can be difficult to see the underlying dimensions that might cut across theoretical orientations and relate to your particular values, preferences, and relational styles. But identifying these dimensions and where you place your self on a dimensional continuum facilitates the personalization of theoretical orientation and effective conceptualization. Table 2.1 provides an introduction and overview of these dimensions.

TABLE 2.1 Dimensions of Focus and Preference for Facilitating Change in Therapy

Contextual Dimensions

Location of Problem	This dimension addresses whether the problem is seen as residing primarily in the individual, in the family, or in the social structural environment, and (relatedly) whether the therapy should focus on changing the individual, family, or social structure or, if this cannot be changed, facilitating different means of negotiating or coping with a problem that is located outside the individual.
Focus of Change	This dimension addresses whether the relative focus of change is on cognition, affect, or behavior.
Past–Present Emphasis	This dimension addresses the relative emphasis on exploring the past or the present in facilitating change.

Relational Dimensions

Directive vs. Non-directive	This dimension addresses the extent to which the therapist directs the content of therapy to facilitate change.
Structured vs. Unstructured	This dimension addresses the extent to which the therapist structures the way in which content is explored (such as through activities, exercises).
Activity Level	This dimension addresses how active the therapist is, for example, how much the therapist talks or shares with the client.
Confrontativeness	This dimension addresses how much the therapist confronts the client. Confrontation is not inherently aggressive, but does relate to the extent to which the therapist calls the client's attention to contradictions, discrepancies, or things of which the client may be unaware.
Significance of Relationship	This dimension addresses the extent to which the therapy relationship is seen as important for the facilitation of change. It includes the centrality of the relationship, which addresses whether the relationship—or the particular nature of the relationship beyond a working alliance—is seen as necessary for facilitating change. If the relationship is seen as central, then this dimension also addresses whether the relationship is seen as directly facilitating or creating change, or is an important means to other mechanisms of change.
Real–Unreal Relationship	This dimension addresses whether the real or transferential relationship is emphasized.
Process Emphasis	This dimension addresses the extent to which the therapist attends to the process dynamics within therapy interactions and the extent that the therapist uses this awareness to facilitate change.

In the next two chapters, we explore these dimensions and related issues of context. In Chapter 3, we explore ecological contexts and introduce the first three dimensions that novice therapists can use to critically evaluate and compare different theoretical orientations, develop their own integrative orientation, conceptualize clients, and formulate treatment plans: location of the problem; focus on cognition, affect, or behavior; and emphasize within the past/present continuum. In Chapter 4, we explore several dimensions related to conceptualizing clients in relationship, including dimensions related to relational style (directive, structured, active, and confrontative) and three dimensions related to conceptualizing the therapy relationship and its role in change.

Exploring and becoming aware of your beliefs and preferences in relation to these dimensions can help you develop a personalized integrative theoretical orientation. These dimensions can help novice therapists consider the reasons why they are drawn to particular aspects of established theoretical orientations. The formal theory associated with traditional orientations frequently places a relative emphasis along each of these dimensions (see Table 2.2). Although each formal theory may generally have a relative emphasis, it is important to remember that there is variability within how therapists adapt or utilize a given orientation. (See Appendix A for a review of the major families of theories.)

These positions are relative and not absolute. For example, a humanistic-existential therapist focuses more on the real relationship than on the transferential relationship and more on the present rather than the past. But that does not mean that he or she is not aware of ways in which given clients may be re-enacting patterns of relating from past relationships. Furthermore, all theoretical approaches believe that their approach creates change in cognition, affect, and behavior (as well as in less individual domains such as relationships), but the theories vary in relation to their initial or primary focus. Furthermore, the placement of each traditional theory in relation to these dimensions is more about how the formal theory describes change rather than a narrow dictate about what therapists do or think about when they identify with that theory. At times, a given dimension is not central to the core elements of the formal theory, in which case there may be variability, as noted in Table 2.2.

These dimensions can help novice therapists address the fit of their own beliefs, values, and styles with established theories of therapy as they work to develop an integrated orientation. These foundational understandings can also help novice therapists consider how their therapeutic orientation affects their case conceptualization and interventions, and when they might want to modify their focus of understanding or their approach. For example, a given therapist may have an integrative theoretical orientation that is primarily psychodynamic and humanistic-existential (emotion focused). These theories focus on the individual in understanding the problem. But if such a therapist was seeing a client who was dealing with sexual harassment at work, they would likely "step outside" their usual theoretical orientation and conceptualize that particular client with a more structural lens, perhaps integrating more of a focus on behavioral change. Of course, it would depend on more fully understanding the client and his or her goals for therapy, but our point is that you can also use these dimensions to consider how to tailor conceptualization to specific clients, in particular contexts.

Tables 2.1 and 2.2 are an introduction to the dimensions and their relation to traditional theory. In the next chapter, we move to a fuller exploration of individuals, with particular attention to the meanings of health and pathology, and the most effective paths to promoting positive change.

TABLE 2.2 Relation of Dimensions to Traditional Theoretical Orientations

	Location of Problem	Focus of Change	Past–Present Emphasis	Directive vs. Non-directive	Structured vs. Unstructured	Activity Level	Confrontativeness	Significance of Relationship	Real–Unreal Relationship	Process Emphasis
Psychodynamic	Individual	Cognitive Affective	Past	Less directive	Semi-structured	Lower	Moderate	Central Means	Unreal	Higher
CBT	Individual	Behavioral Cognitive	Present	More directive	Structured	Higher	Moderate to higher	Not Central	Real	Lower
Humanistic Client-Centered	Individual	Affective	Varies	Less directive	Unstructured	Lower	Lower	Central End	Real	Moderate
Humanistic–Existential	Individual	Affective	Present	Varies	Structured	Varies	Moderate to higher	Central Means	Real	Higher
Systems-Ecological	Family	Varies	Both, greater present emphasis	Varies	Structured	Moderate	Varies	Central Means	Real and Unreal	Moderate to higher
Constructivist	Varies	Varies	Both/varies	More directive	Varies	Moderate	Varies	Central Means	Real	Moderate to higher
Liberation	Structural	Varies	Both, greater present emphasis	Somewhat directive	Varies	Moderate	Moderate	Central Means	Real	Moderate to higher

© Cengage Learning 2013

SUMMARY

We focus in this chapter on exploring the values and worldviews that influence your choice of elements of theoretical orientation and your case conceptualization. We ask you to explore your beliefs about how change happens, and how your own process of change has affected the development of these beliefs. We discuss influences on your understanding of change, related to your cultured worldview as well as your experiences with power, privilege, and oppression. We relate your beliefs about change to an exploration of your epistemological philosophy, which also influences your understanding of health and pathology. Finally, we present an overview of several dimensions of change that can facilitate personalization of theoretical orientation and effective conceptualization that we will explore in the next two chapters.

REFERENCES

APA Council of Representatives (2011). *Guidelines for psychological practice with lesbian, gay, and bisexual clients*. Washington, D.C.: American Psychological Association.

Diagnostic and Statistical Manual of Mental Disorders, Fourth Edition (DSM-IV). Washington, D.C.: American Psychiatric Association.

International Statistical Classification of Diseases and Related Health Problems (ICD-10). (2010). World Health Organization.

Markus, H.R. & Kitayama, S. (1991). Culture and the self: Implications for cognition, emotion, and motivation. *Psychological Review* 98, 224–253.

Nagata, D.K. (1993). *Legacy of injustice: Exploring the cross-generational impact of the Japanese American internment*. New York, N.Y.: Plenum Press.

Norcross, J.C. & Prochaska, J.O. (1983). Clinicians' theoretical orientations: Selection, utilization, and efficacy. *Professional Psychology: Research and Practice* 14, 197–208.

Sue, D.W. & Sue, D. (2007). *Counseling the culturally diverse: Theory and practice* (5th ed.). New York, N.Y.: Wiley.

Vasquez, H.M. & McGraw S. (2005). Building relationships across privilege: Becoming an ally in the therapeutic relationship, pp. 61–43. In M. Mirkin, K.L. Suyemoto, & B. Okun (Eds.). *Psychotherapy with women: Exploring diverse contexts and identities*. New York, N.Y.: Guilford Press.

Wampold, B.E. (2001). *The great psychotherapy debate: Models, methods, and findings*. Mahwah, NJ: Laurence Erlbaum.

3

Conceptualizing Clients in Contexts

In this chapter, we focus on how you understand different contexts that affect your conceptualization of clients' experiences and their needs for psychotherapy. From an ecological framework, we understand clients' experiences to be shaped by the contexts in which they currently live and those in which they have developed. In this chapter, we explore the ecological model and particular contexts that influence clients. We then apply this general understanding of contextual influences to developing an understanding of our theoretical orientation and influences on case conceptualization for particular clients.

EXPLORING CONTEXTS

Every client is a unique individual with personal temperament, strengths, weaknesses, likes, and dislikes. But all clients also exist in social and environmental contexts that affect how their unique individual qualities are experienced and expressed. There are many different ways to understand people in contexts. Ecological models work to consider not only the internal experiences and choices of an individual, but also the reciprocal relationships with people and with environments (things, places) and ideas associated with environments (such as regional cultures). Some ecological approaches emphasize environmental systems and the relationships and interactions between them, such as

Bronfenbrenner's Ecological Systems Theory (1979). Other approaches emphasize relationships, considering different layers of intimacy and connectedness (Okun & Kantrowitz, 2007). Still other approaches emphasize identities and the ways individuals understand themselves influenced by different ecological levels (Root, 1998). Other approaches emphasize the levels of shared identities that people have based on their similarities and differences (Sue & Sue, 2003). All of these approaches attempt to consider what kinds of relationships, environments, experiences, and identities influence and are influenced by individuals.

Table 3.1 and Figure 3.1 illustrate our approach to considering multiple contexts and different types of influences within these contexts that are helpful when exploring how contexts affect individuals. Considering the influences of ecological contexts for individuals is complex because the influence of a given context is related to multiple types of experiences, including:

- The relationships that people have, which encompasses the kinds of people with whom one has relationships and the nature of those relationships (close, distant, conflictual). How do these relationships affect people?
- The environmental context, which encompasses the physical aspects of the environment including the things, people, events, or resources in the environment. Does a family live in a large house or a small apartment? Is the neighborhood urban or rural? Who lives in the neighborhood? Who attends the organization? For example, are working class women of color present at the feminist social organization?
- The ideologies, which encompasses the beliefs and values characterizing the context. What are the roles and rules for being a member of the family, a student at a particular school, a good neighbor, a member of the Asian American racial group? Another way to approach this is to consider the feeling of a context. Is the work setting generally collaborative or competitive? Is the region welcoming to Gay/Lesbian/Bisexual (GLB) relationships and identities (Is same-sex marriage permitted)? It is important to remember that just because the majority of people in a context share a particular ideology, that does not mean that any given individual will share all (or any) of those beliefs and values. Furthermore, most individuals personalize ideological systems, such as a Roman Catholic who believes in abortion. But the individual may be affected by the context's ideology regardless of his or her agreement with it.
- The identities, which encompass the ways in which people make meaning of their relations to their contexts, the salience of those contexts to their self-concept, and the particular kinds of self-referencing that are important to them.

Each of these four aspects interacts with the others and the context, as a whole, affects individuals. For example, the ideology of a family affect the kinds of relationships that are valued and enacted, the environment that the family creates or chooses, and the kinds and nature of identities that are encouraged or discouraged within families. Simultaneously, the environment

TABLE 3.1 Ecological Contexts and Influences

Context	Relationships	Environmental Aspects	Ideology and Practices	Related Identities
Family and intimate relationships	Family interactions and relationships	Family/home: location, setting, family make-up	Gender roles, family rules, authority, communication patterns, culture as it affects family	Daughter, son, father, family member
Extrafamilial relationships	Extrafamilial relationships such as friends, colleagues	School, work, organizations, environments for recreation and hobbies, shared projects	Ideas about achievement at school, meanings of being a friend, value of activities such as hobbies)	Student, church member, artist, career identities
Neighborhood and community	Acquaintances, shop people	Neighborhood and community	Community value on education, political values	Community member, neighborhood member
Sociocultural and sociostructural systems	Social referent groups (culture, religion, sexual orientation, gender, nationality)	Location, (region, country), group location and environment (ethnic enclaves, group-specific organizations)	Beliefs about a higher power, communication patterns, attitudes towards other racial groups	Racial identities, ethnic identities, regional identities
Universal	Humanity	World		Human

© Cengage Learning 2013

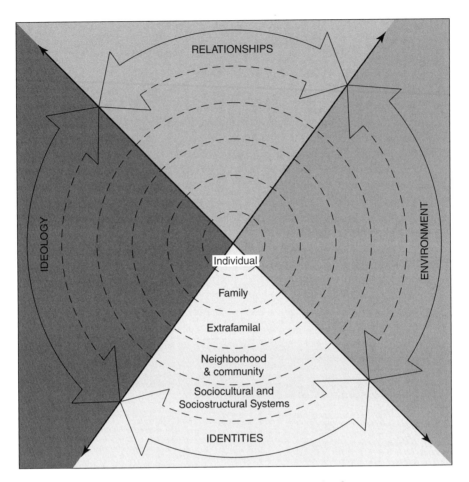

FIGURE 3.1 Schematic of Interacting Ecological Contexts and Influences
Adapted from Bronfenbrenner (1979), Knopf (1986), and Okun and Kantrowitz (2007).

affects relationships, ideology, and identities: an individual living in a very rural area where houses are far apart will have less daily social interaction, be more likely to value or enjoy country life, and be less likely to identify as an "urbanite" than someone living in Manhattan.

In Table 3.1, we offer examples of each type of influence within each identified context, but it is important to remember that there are many other possibilities as well. In addition, it is important to remember that every individual will experience the particular constellation of context influences and interactions among contexts differently because the experience is also affected by unique individual experiences and interpretations. And the importance of different contexts and influences within contexts will also vary. The meanings of most of the contexts are well known, but our meaning of "sociocultural and sociostructural systems" may need clarification. Sociocultural and sociostructural systems are inherently interrelated contexts that are related to social reference groups (sociocultural), and to the power and privilege that shapes the boundaries, salience, and experience of those reference groups (sociostructural). The sociocultural emphasizes the cultural aspects of this, focusing more on the meanings shaped by individuals and groups themselves, while the sociostructural aspect focuses more on the influence of power relations, including current and historical experiences of hierarchy, oppression, and privilege.

In each context, an individual may function differently with regard to role, rules, function, and responsibilities. For example, a client may be highly respected and high functioning in his work role, but withdrawn and passive with his wife and children; sociable with his golf partners, but unsociable with family friends and neighbors. Each context brings out different aspects of a person; there may be some persistent traits and characteristics, but there is not necessarily a unified perspective about an individual that is completely consistent across contexts.

EXERCISE 3.1 ▪ Review Table 3.1 and Figure 3.1. Consider your own experiences in relation to each of the contexts described, exploring your experience of each context in relation to your relationships, and the particular environment, ideologies, and identities that are important to you. How have these contexts shaped you as a person? Imagine that one or more of these contexts or the aspects of experiences within them were very different from what it has been in your actual experience. How would you be different?

EXERCISE 3.2 ▪ Using Figure 3.1, discuss with classmates in a small group how you function in different contexts as you try to identify significant influences in each context contributing to your choice of career. Try to focus on your roles, strengths, and weaknesses in each different context.

EXERCISE 3.3 ▪ What do you know about Nancy in each of these contexts? Draw a circle for each context in which Nancy lives: family, school, peer group, and so on. What questions do you have about Nancy's different contexts? How might different answers affect your conceptualization of her? What are her possible roles, strengths, weaknesses in each different context? Consider her developmental stage: for example, being a college student, identity seeking, emotional separation from family. Which context(s) might be a top priority to address at this time? Discuss your ideas with others in a small group.

EXERCISE 3.4 ▪ Consider the similarities and differences between Nancy's contexts and your own. What biases or preconceptions about these contexts might affect your conceptualization? For example, how is Nancy's experience with her peer group similar or different to your experience?

While we believe that all of these contexts influence people, we also believe that most people are affected most by a) the family context because it is most often the context in which they have their closest relationships, and b) the sociocultural and sociostructural systems context because the other contexts are so strongly related to being nested within this context. We will, therefore, focus our discussion in the remainder of the chapter on these two contexts, while acknowledging their interactions with other contexts and the fact that our beliefs may not apply to every individual.

Exploring Family Contexts

We believe strongly that individuals cannot be viewed or treated without consideration of the context of their most significant relationship systems, primarily their family of origin and their current family or intimate relationships. Within the client's family system, there are different subsystems—the marital, parenting, sibling, female, male, older, younger groups in which the client may experience him or herself differently than with the larger family system. There is a basic reciprocal connection between the individual and his or her relational contexts; it is not unusual for an individual's symptoms to not only maintain the relationship system, but also be simultaneously maintained and stabilized by it. For example, if your client grew up in a family where men are supposed to be the "providers," and women are supposed to stay home and socialize, she may have difficulty maintaining empathy and support for her husband who has been laid off and whose unemployment insurance is running out. And she might resent the suggestions from other people that she try to get a job. The attitude and behavior of this client might stress the couple relationship.

Learning about and understanding the relationship roles, operating rules, communication processes, and interactive patterns in a client's family of origin provides valuable clues to understanding the client's current relationship functioning and the meanings of his or her thoughts, behaviors, and feelings. Equally important is understanding the boundaries (physical and emotional) in the client's family of origin: Who was close to whom? Were feelings able to be expressed? How were decisions made? Were the boundaries rigid? Permeable? Disconnected? There is much literature on family systems (see Nichols & Schwartz, 2006 for a summary) and their pervasive influence on individual members. We want to learn as much as we can about the family's cultural style in order to understand fully the beliefs, values, and attitudes with which the client grew up as well as the multigenerational ideological beliefs and relationship patterns. Bowen's (1978) transgenerational theory considers not only the present family presentation and relation but also the way that the family enacts historical patterns through the multigenerational transmission of relationship patterns in family systems. We typically ask clients to tell us what it was like growing up in their family; how feelings were permitted and expressed; how conflict was dealt with; who was close to whom; how you got what you needed and wanted, and so forth. Some of this information might be formatted on a genogram. A family genogram diagrams at least three generations of the family and utilizes symbols or written notes to record ages, divorce, death and cause of death, occupation, critical incidents, and so forth. (See Appendix C for Nancy's genogram.)

EXERCISE 3.5 ▪ What are some of the roles and rules in your family? How have these roles and rules affected your relationships with other people? What is your understanding about what is good or problematic in your own experience? What were your roles as a child and a sibling? Was your child role different with each parent and with your parents as a couple? If you grew up in a blended family or in multiple families, how did that situation affect the roles and the rules and the way you were affected by them? Do you think the roles and rules were adaptive as you grew older?

The individual client's story about him or herself, about his or her family, and other relationships is one perspective. But whenever possible, it is helpful to have at least one session with the individual client and significant others as well and to obtain information (ethically and legally) from other current and former treatment providers, teachers, and so forth. Everyone has a different "story" and family members have both shared and non-shared experiences. For example, a family session with Nancy and her family revealed that her sister did not agree with Nancy's perceptions of what it was like to live in their home.

EXERCISE 3.6 ▪ Discuss in small groups the following case about Ned Ned, age 22, withdrew from middle school in the eighth grade and was home-schooled for four years. At the time of his withdrawal from school, he was thought to be school phobic and to suffer from "anxiety." He had no treatment and self-referred to therapy at the age of 20, saying he needed to "get his life together." He refused to take medication of any sort and required his mother to drive him to therapy as he could not take the time to utilize public transportation. His mother reports that she and he fight continuously, that he refuses to talk to his father, and that all he does is play computer games all day at home. Ned lives with his parents and his younger sister in a small suburban town that is primarily White middle-class Catholic. Both parents have advanced science degrees and met in graduate school; the mother stopped working when Ned was born and has needed to be with him every minute since then. In two family sessions, the therapist learned that: 1) the family consider Ned as being a little "anxious" and needing time to grow up; 2) when he left school at age 12, he did not want Treatment, so they did not pursue this and it never occurred to them to talk to anyone to try and find out what was going on; 3) the family's only outside relationships are with extended family; 4) the family highly values education and expect Ned to use his "intellectual gifts" and go to college. Ned had been admitted to a college, but could not leave home even to take classes at a local college because it would involve taking public transportation. His family did not challenge him on this. How would you conceptualize the family context given the information? How might family sessions help you to understand Ned? Which of your feelings might influence how you think about and work with him?

EXERCISE 3.7 ▪ Consider the following interaction that occurred in a family session with Nancy: In this session, you observe a heated exchange between Nancy and her sister over entry into each other's room and taking the other's clothing. As the anger between the two sisters escalates, you sit back and watch mother tell Nancy to stop fighting with her sister as she puts a protective arm around her younger daughter. The father just rolls his eyes and lets the mother take charge of the situation. What meaning do you make of this family interaction? How does it make you feel? How does it help you to understand Nancy? Which of your feelings might influence how you think about and work with Nancy?

It is frequently not enough to know about general family structure or dynamics. A therapist may also need to know about the particular family developmental stage or constellation for a given client. This is particularly important for therapists who practice with a relative emphasis on the family context, rather than on the individual. But even with an individual focus, the therapist should be knowledgeable enough about the specific context to be able to conceptualize the client's unique experiences and perspective in relation to the modal experience. For example, if you are seeing a client who is pregnant with her first child, it will be important for you to know about this family stage and the ways that this experience both influences and changes family dynamics and issues.

Exploring Sociocultural and Sociostructural Systems

The sociocultural and sociostructural systems include two interacting aspects of context: 1) the aspect of social system context that is cultural and 2) the aspect of social system context that is related to power and privilege.

Cultural Contexts: The Sociocultural Aspect Having a shared culture means having beliefs, values, behavioral norms, and traditions associated with the group and transmitted within the group. Just as it is important to understand your own cultural context, it is equally important to understand the cultural contexts of your client. Understanding the cultural contexts of the client involves not only knowing that diversity exists and being open to learning about the client's specific cultural contexts, but also knowing about the diverse values, behaviors, and relational norms for different groups.

There is a vast amount of research that explores diverse cultural experiences and describes important inter- and intra-group variability related to ethnicity, social class, gender, sexual orientation, and other social group variables that are salient in the United States. Books such as Sue and Sue's *Counseling the Culturally Diverse* (2007) provide overviews of this research. If you are seeing a client from a cultural background different than your own, it is important that you seek out and educate yourself about his or her cultural background, so you can conceptualize his or her experiences and current challenges within the appropriate referential space. The client should not have to educate you about what African Americans "are like" as a group, or about the elements of gay culture overall.

EXERCISE 3.8 ■ What do you know about cultural differences between racialized pan-ethnic groups related to the following values? Choose a specific ethnic or pan-ethnic group (such as Latinos) and do some research into the modal experience of the group you have chosen. Consider the table of cultural values below.

Individualism	Compliance	Spirituality
Conformity	Freedom	Improvisation
Communication patterns	Achievement	Family
Obligations	Support to elders	Collectivism
Cooperation	Competition	Loyalty
Engaging or avoiding conflict	Harmony	Time orientation
Challenging authority	Gender roles	Interdependence

What do you think the implications of these differing cultural values are for the ways people negotiate relationships and life issues? Are the differences inter- or intra-cultural?

It is important to know about the general importance of culture and the various components of cultural worldviews and experiences that make up the sociocultural aspect of context. It is also important to know about the modal experiences of broad groups such as the major pan-ethnic racialized groups of White Americans, Black Americans, Latinos, Asian Americans, and Native Americans that are often discussed in texts on multicultural counseling. But it is equally important to consider the heterogeneity within these broad groups. Just as a therapist needs to understand the particular family context and development for a given client, a therapist needs to know about both the general culture of a particular client and the particular ways the client experiences and interprets cultural variables.

For example, if you are seeing a client who is Mexican American, it will be important for you to have some knowledge about Mexican American culture and community specifically and not just about the importance of culture or about Latinos in general. Having this kind of knowledge is important for several reasons. First and foremost, it enables you to conceptualize health and pathology, treatment goals, and evidence based practice within the client's particular context. It also communicates expertise and care to the client. Finally, it enables you to consider and utilize additional resources for the therapy or the client directly.

EXERCISE 3.9 ▪ Mark, a White European American Protestant therapist, is seeing a client, Rachel, to address issues related to anxiety, particularly social anxiety in larger groups. Rachel is also a White European American, but she identifies with Conservative Judaism. Rachel has mentioned in passing (during discussions of behavioral homework) her frustration with her graduate school program, which regularly schedules social events on Friday evenings. Mark has not had a lot of exposure to Judaism generally and the friends he has that are Jewish are Reform Jews, with whom he frequently socializes on Friday evenings. However, he is aware that the Jewish Sabbath begins on Friday evening at sundown, and that some Jews observe the Sabbath differently than his friends, so he has sympathized with Rachel about her frustration. Do you think that Mark should know more about Judaism to work well with Rachel? Why or why not? Are there particular instances or issues that you feel would demand that he know more about Judaism generally, or about Conservative Judaism specifically?

As it happened, Mark needed to reschedule some of his appointments with Rachel during the months of September and October. There are many Jewish holidays in September and October, including Rosh Hashanah, Yom Kippur, and Sukkot. When offering dates to reschedule with Rachel, Mark included dates that conflicted with some of these holidays. How do you think Rachel felt about this? What effect might it have had on the therapy? Mark had been seeing Rachel for a few months when this happened, and Rachel was able to share her feelings of disappointment and alienation with Mark. But this particular issue "derailed" their focus on Rachel's social anxiety and was more of a learning opportunity for Mark than a therapeutic moment for Rachel. It would likely have been more therapeutic for Rachel if Mark had been aware of these holidays and their meaning and observance for Conservative Jews.

How does one learn about particular contexts? Your training cannot include information about all of the specific identities, groups, and subcultures that will be reflected in your clients' experiences. As mentioned above, most training handles these issues broadly, for example, discussing African American cultural norms with little attention to other cultural groups within the Black racialized group or to regional differences. But the experience of Sudanese refugees or Haitian immigrants may be quite different than that of African Americans, and the experience of being Black in Detroit is not the same as the experience of being Black in Honolulu, New York, Memphis, San Francisco, New Orleans, or Oglala South Dakota. It will frequently be up to you to seek out information within your particular regional context and for each client's particular contexts. Some strategies for increasing your knowledge and experience include:

- Reading literature within the field. You can always do an online literature search on PsycInfo or a similar database. There are many published articles and books that present overviews of information related to therapeutic

work with clients from particular cultures, family constellations and stages, or other contexts (see McGoldrick et al., 2005).
- Use the Internet. As an overview source, the Internet provides a wealth of basic information.
- Search for regional community events and resources. If possible, it is important that you have more than "book knowledge." You may not have the time to do this for all kinds of diversity or for all clients, but this is particularly important if you are seeing several clients from similar cultures or contexts. Attending community events is one way to get a more experiential sense of a context. It is also helpful to be aware of resources that exist in your area. You can contact the organizations or individuals providing those resources (a community based organization providing legal resources for undocumented immigrants, a social group for new mothers, or a support group for individuals caring for dying parents). These contacts can familiarize you with the resource and also provide you with additional information about the context from the "expert" point of view, including the specific regional issues. They can also frequently point you to additional resources and community/contextual supports. You can also ask if you can visit the organization or group.
- Foster relations with diverse people in your personal and professional life. They can help you understand contexts different than your own, point you to additional resources, and act as consultants in moments where there may be a disconnection between the clients' experience and your knowledge.

At the same time as we need to be familiar with cultural diversity of different groups, it is necessary to understand the client's particular relationship to his or her sociocultural systems contexts(s). There is an important distinction between the modal experience related to a cultural affiliation and the individual manifestation of a cultural affiliation (Okun, Fried, & Okun, 1999; Constantine & Sue, 2006). The modal experience is what is characteristic about a cultural group as a whole. For example, Asian Americans as a group are less emotionally expressive and more interdependent than European Americans as a group (see review in Uba, 1994); women as a group provide more social support than men as a group and are judged differently when they are assertive (see reviews in Crawford & Unger, 2004); African Americans as a group are more suspicious and less trusting than European Americans (see reviews in Sue & Sue, 2007). Although these descriptions may be true of the group as a whole and in comparison to the dominant group, each individual within the group has unique experiences. For example, some Asian Americans may be very emotionally expressive and more independent, and some European Americans may be less emotionally expressive.

Also, each individual relates differently to the modal experience of various characteristics of their reference group. A particular woman may be both highly socially supportive and highly assertive, reflecting the modal experience of women as a group with the first characteristic (social support) and differing from the modal experience of the group with the second characteristic (assertiveness). Individual manifestation of cultural affiliations may also vary within a person across contexts. Furthermore, individuals may have different salient or intersecting sociocultural influences: at a meeting one of us attended, an African American woman surgeon told a group of African American male physicians that her gender caused her more difficulties than her race in professional contexts, which truly shocked her male colleagues! Intersectionality theory evolved from feminist sociological theory (Crenshaw et al., 1989) in order to study the multiple, simultaneous influence of oppressive systems on social relations. Different identities and statuses will interact to shape the particular experience of individuals.

EXERCISE 3.10 ■ Interview at least two people who are members of the group you chose in Exercise 3.8 about their particular experiences and relation to their ethnic cultural group. Consider how individuals within a culture may vary. What do these exercises tell you about the importance of understanding both the modal and individual experience of culture?

Our knowledge of the modal experiences of cultural groups is like a very tentative group of hypotheses that we need to hold lightly, always remaining open to constant modification. It is important to know about modal experiences and the ways that cultures and status variables shape individuals because we always begin our conceptualizations with initial hypotheses. If we do not have understanding of diverse cultures, then we are likely to begin with hypotheses that reflect the dominant cultural group (the White European American male group in the United States) or our own experiences. We want to begin with our best approximation of where the client is beginning, and understanding group diversity helps us do this. But we must use our knowledge of cultural norms and variability carefully, avoiding a cookbook or prescriptive approach. Although it is not the client's responsibility to educate us about the general experience of the group, it is the client's responsibility to communicate his or her particular experience as a member of a particular group. And it is our responsibility to seek out these particularities in order to modify our initial hypotheses.

Sociocultural Systems As Related to Power and Privilege: The Sociostructural Aspect We frequently think about culture in relation to ethnicity, but there are cultural aspects to being Muslim (a religious identity), gay, upper class, and many others. There are also cultural aspects to being Republican, or punk, or a banker. But these latter categories are not defined by their relation to structural power and privilege. What sets sociocultural systems related to power and privilege apart from culture generally and inherently relates them to the sociostructural aspect is that the meanings of these contexts are not primarily determined by the group itself, but by meanings in a larger social system. Some social categories have effects on individuals regardless of their choices or behaviors and regardless of whether they maintain relationships, frequent environments, endorse ideologies, or claim identities related to those contexts. These kinds of context effects are frequently *imposed social constructions* on individuals or related to individuals' reactions to what is being imposed.

As we discussed in Chapter 2, there are some group or identity statuses that are related to having more or less privilege in the context of the United States. Given that this is so, one aspect of understanding the client's sociocultural systems context is also being familiar with the issues of power and oppression that shape the experiences of individuals and groups. This is particularly important when the client is a member of an oppressed group within the context, as it is important that the experience of therapy is not an experience that reifies the client's experience of oppression. But it is also true when the client is a member of the dominant group, although the client may have less awareness of the influence of the context.

Ethnic culture is, itself, a sociostructural system because there is a dominant ethnic culture in the United States, and those from "minority" ethnic cultures are affected not only by their own culture, but by the dominant culture and the dominant culture's judgment of their own culture. Because of this, it can be difficult to understand whether an individual is being affected simply by cultural contexts and their incorporation into individual and family experiences and worldviews, or by the sociostructural aspect of that system. For example, Laura Uba (1994) describes how Asian Americans are more likely to experience social anxiety than European Americans. A common way to understand this difference is related to the sociocultural aspect that most Asian cultures emphasize high context communication and sensitivity to social cues. But Uba questions whether this difference may also (or alternatively) be due to experiences that Asian Americans have had individually or historically as a group with exclusion or rejection due to ethnocentrism or to racism.

In many cases, the cultural meaning and the effects of the socio*structural* system simply cannot be separated. Remember, by sociostructural system, we mean the historical experiences and the legacies of those experiences that relate to power and privilege. For example, given the history of Native Americans in the United States, it is impossible to know what Native American culture (relationships, environment, ideology and practices, identities) might be like without the influence of extreme oppression characterized by genocide and active deculturization such as the practice of taking Native American children from their families and communities and placing them in boarding schools where they were beaten for speaking in native languages or practicing cultural traditions. The meaning of being Native American is inherently related to the history of the group's oppression.

Individuals in the dominant status in relation to sociostructural systems are also strongly affected by these systems, although they may be less likely to be aware of this influence. People in dominant statuses frequently feel more entitled to judge others and impose restrictions on others that they, themselves, would find unacceptable. While sometimes these effects are more blatant (such as when they are exhibited in individual acts of discrimination), more frequently these effects are much more subtle. For example, men's communication patterns are characterized by interrupting and speaking more (Crawford & Unger, 2004; Tannen, 1996). These patterns are thought to be shaped by being in the dominant group, feeling entitled to have one's voice heard, and having an unconscious assumption that one's own contributions are most vitally important. However, many men actively embrace social justice and equality for women; they do not *intend* to be affected by being more privileged within the sociocultural system.

Other sociocultural systems similarly have complex intersections of culture, power, and privilege (sociostructural aspects), particularly in relation to a dominant group. For example, lesbian culture may be characterized by less rigid gender roles, closer intimacy among partners, and a greater emphasis on harmony or interdependence. A lesbian may be influenced by this culture and also by the sociostructural (rather than cultural) meaning of being a lesbian, including experiencing homophobia, feeling pressure to stay "in the closet," exposure to stereotypes of lesbians, job or housing discrimination, and barriers to establishing recognized intimate relationships and family.

EXERCISE 3.11 ▪ In small groups, discuss how you think that individuals may be affected by experiences of discrimination. Consider the effects on the individual, the relationships, environmental aspects, ideology, and identities (see Table 4.1). Consider also the effects of discrimination in relation to different contexts: discrimination within the family, within extrafamilial relationships like work colleagues, within the neighborhood such as from shopkeepers, or within a general social context such as in the media. Consider how discrimination based on race, sexual orientation, gender, social class, religion, or other factors may be similar or different in their effects. How will your own experiences with discrimination influence your understanding of clients with similar or different experiences?

EXERCISE 3.12 ▪ Nancy is in a socially privileged space in relation to race (White), ethnicity (European American), social class (upper middle class), and religion (Christian), and a socially marginalized space in relation to gender (female). How might her experiences with privilege or oppression affect her?

DIMENSIONS RELATED TO CONTEXTUAL UNDERSTANDINGS

Case conceptualization is about how you understand the client, yourself in relation to the client, and the space between you—all in multiple contexts. In approaching the case conceptualization of a client, we want to explore these different areas as thoroughly as possible, considering multiple possibilities and perspectives. We also need to consider how our own emphases on the influences of different contexts affect our conceptualization of clients. To begin this process, we introduce three dimensions of focus and preference for facilitating change in therapy: 1) context location of the problem, and two dimensions related to focusing within context; 2) focus of change; and 3) past/present emphasis.

Context Location of the Problem

People vary in how much they attend to the influence of different contexts on their experiences. Some people think a lot about how their family affects them, while others are focused more on personal individual choices or characteristics. This relative emphasis is affected by our values and social experiences, by our developmental life stage, as well as by the contexts themselves. In relation to sociocultural and sociostructural systems contexts, people who are more constrained by the latter (such as those who experience discrimination) are likely to pay more attention to the effects of sociocultural and sociostructural systems contexts and statuses than those who are less constrained. For example, people of color are usually more aware of race as an influencing factor than White people. That does not mean that race is most salient all of the time; in some contexts, gender or generation may be more salient than race as noted in the case of the Black female physician above, or family or personal levels may be most salient.

Similarly, therapists vary in their view of the extent to which different contexts affect the development of psychological experience. It is important to address the therapist's understanding of how experiences or problems develop and, thus, what needs to be and/or can be changed. Problems may be conceptualized in relation to the individual, within relational (such as family) contexts, or within sociocultural/sociostructural systems contexts. Of course, most experiences are influenced by all three of these levels (and the other contexts described above), so it is important to realize that we are talking about relative emphases, not simple categorical choices. Therapists tend to vary in how much they emphasize each level or which level they consider first which, in turn, influences and interacts with the therapists' orientation.

EXERCISE 3.13 ▪ Based on the information you have so far about Nancy, write the following lists: 1) What five things do you most want to know more about Nancy in order to help her?; 2) How would knowing these things help you consider effective interventions for Nancy?; 3) What problems or issues are most important for Nancy?; 4) What do you think might underlie her presenting concerns?; 5) What are your first thoughts about how you might address these concerns? Try to generate as many diverse ideas as possible.

The characteristics of different clients and their contexts will also affect the extent to which a therapist attends to different contexts in conceptualizing clients' problems. Some clients' problems will seem to be more directly related to individual, family, or sociocultural systems contexts than the problems of other clients.

Consider another client:

> Thanh is the 15-year-old son of a Vietnamese refugee family living in a suburban area with significant Black, White, and Vietnamese populations. He lives at home with his parents, older sister, and uncle. Thanh's mother and uncle work long hours in a restaurant and have done so all of Thanh's life. Thanh's father does not work; he is reclusive and withdrawn and Thanh has heard him weeping and yelling late at night. Thanh's two older brothers died on the refugee journey before Thanh was born and his older sister is currently attending a community college and working as a waitress. Thanh was referred to you by his high school counselor because of repeated fighting at school. Recently, the school has had multiple problems with fighting, particularly between Black and Vietnamese students, and between Vietnamese and White students. However, Thanh's issues with fighting seem to pre-date these major incidents. The school describes Thanh as angry and oppositional, and his school counselor expresses concern that Thanh is becoming involved with a Vietnamese youth gang.

EXERCISE 3.14 ■ Based on this information about Thanh, write the following lists: 1) What five things do you most want to know more about Thanh in order to help him?; 2) How would knowing these things help you consider effective interventions for Thanh?; 3) What problems or issues are most important for Thanh?; 4) What do you think might underlie his presenting concerns?; 5) What are your first thoughts about how you might address these concerns? Try to generate as many diverse ideas as possible.

Therapists who emphasize the individual in their conceptualization will be more likely to consider first (but not only) issues related to the person of the client. What kind of personality does the person have? What are his or her thoughts or feelings about the situation? What motivations does he or she understand? Therapists who emphasize the individual do not disconnect the individual from family or intimate relationships and sociocultural systems, but are most likely to see relationships and sociocultural systems as contexts within which the individual resides. These contexts may influence the individual, but it is the individual's perception of and reaction to these contexts that lead to individual choices related to health or pathology.

A therapist who emphasizes the individual would have questions about Nancy's "difficult" personality and how her self-centeredness creates conflicts with other people. He or she might wonder whether Nancy's feelings of isolation and deprivation relate to a view of herself as unlovable. This therapist might view her anger with others as indicative of poor impulse control. The focus of treatment might be on changing her attitudes, managing her anger, exploring her self-concept, and becoming more flexible and adaptive.

A therapist who emphasizes the individual may have many questions about Thanh's interpretations of events and about his personal tendencies or personality. They may question whether Thanh is impulsive, angry at his parents or his situation, alienated from his family or his culture, has poor anger management, and possesses low self-esteem such that fighting is a way to feel powerful or positive. This therapist might work with Thanh to better manage his anger, to explore his feelings about his family and his peers, or to change his negative thoughts and perceptions about how others treat him.

Therapists who emphasize the relational system in their conceptualization will be more likely to consider first (but not only) issues related to the family or to other important relationships of the client. How do the expectations of others shape the thoughts, feelings, or behaviors of the client? What roles is he or she expected to assume? What relational purpose do the client's problems serve? Therapists who emphasize relational systems do not discount the individual or the sociocultural/sociostructural context. They recognize that the individual makes choices about how he or she interacts with relational systems and that relational systems are shaped by sociocultural/sociostructural contexts. However, they see the relational systems as primary influences on thoughts, feelings, and behaviors. Such therapists see problems as related to the interactive relationships between people, not primarily the individual choices or the larger sociocultural or sociostructural processes or constraints.

A therapist who emphasizes relationships would have questions about how Nancy became the "scapegoat" in her family, what her family's expectations were of her, and how they manage conflict within the family. This therapist would want to understand the family structure, the flexibility/rigidity of roles, and the rules and patterns of communication. He or she would also want to know about other significant people in Nancy's life: her friends, classmates, and relationships with teachers. The belief would be that, while Nancy may have a "difficult" temperament, family and other relational influences shaped her development and the maintenance of her behavior.

A therapist who emphasizes relational systems may have many questions about Thanh's relationships with his family, about his parents' relationship with each other, and about the family history. He or she might hypothesize that Thanh's fighting deflects attention from the problems within the family or that Thanh's fighting reflects his reaction to witnessing his father's feelings of helplessness, which Thanh may or may not relate to the legacies of war trauma. This therapist might work with Thanh's family to change the roles and expectations, not only of Thanh but also of other family members, as Thanh's experience is seen as inherently related to the experiences of those with whom he has close relationships.

Therapists who emphasize the sociocultural/sociostructural system in their conceptualization will be more likely to consider first (but not only) issues related to sociocultural experiences and group affiliations related to power and privilege and sociostructural roles and constraints. What sociocultural and sociostructural statuses and groups does the client belong to? How do these affiliations affect the client? What are the social and institutional experiences and constraints related to being part of these groups? Issues of oppression and constraint are particularly important influences on mental health and psychological problems.

A therapist who emphasizes sociocultural systems would want to know about the sociocultural values of Nancy's community. How has living in an affluent community shaped her values and perceived needs? How did it feel to belong to a family less affluent than the families of most of her peers? Such a therapist would want to understand Nancy's sociostructural experiences with power and privilege. What ideas of privilege did growing up in this community engender, and how did her family relate to others in the community? What messages came from the family and from the larger community about the way things are supposed to be for Nancy? The therapist would also recognize that much of Nancy's self-esteem came from being a good student in a competitive academic community school.

A therapist who emphasizes sociocultural systems may have many questions about how events related to sociocultural and sociostructural systems have shaped Thanh and his family, such as the trauma of being a refugee, the experience of adjusting to a new culture, and possible racism. He or she might work with Thanh to understand how the Vietnam War and the refugee experience may have affected his parents and created distance in the family and help him become aware of racialized dynamics among his peers. This therapist might also intervene with the school system to address race relations in a more systematic way, or help Thanh to do so.

We continue to emphasize that good therapists attend to all three of these levels, as well as the many complicated interactions between them and the other contexts described above. Most therapists, however, place a relative emphasis on one or two of these levels, considering the others as influences rather than primary areas for intervention.

EXERCISE 3.15 ■ Go back to the earlier questions regarding Nancy and Thanh in Exercises 3.13 and 3.14. What kinds of questions were you asking? What does your list tell you about your relative emphasis of these three levels? For each problem, consider whether the problem or particular aspects reside within Nancy and Thanh as individuals, within relationships (which ones), and/or within larger sociocultural systems.

As we have noted before, the particular presentation of a client may relate to the extent to which a therapist attends to different contexts in conceptualization. But we need to be wary of only attending to some contexts with some clients or some presenting problems. Just as individual and family influences affect all clients, sociocultural and sociostructural contexts influence all clients, not just "diverse" clients (or those that are not part of the dominant group).

For example, a study by researchers at Boston College elucidates the psychological challenges and relationship constraints (Do people like me because I have wealth?) with which wealthy individuals and families contend and debunks many of the social myths that we have about "money buying happiness" Schervish (2011).

Relation of Context Emphasis Dimension to Etablished Theories
Psychodynamic, phenomenological, cognitive, and cognitive behavioral psychotherapy theories are most consistent with emphasizing the individual rather than the contexts. These orientations more frequently view the problem (and solution) residing within the individual, although there is some variability; existential and gestalt therapists may place a secondary emphasis on sociostructural contexts, for example. These theories tend to utilize direct interventions to change individuals.

In contrast, family systems theories are most consistent with emphasizing the family context, locating the problem(s) within relational systems. Family systems therapists tend to utilize interventions directed towards changing the dynamics of relational systems. Feminist and multicultural theories attend primarily to sociostructural systems, actively considering how systems of oppression and privilege shape the development of psychological experiences and problems. Feminist and multicultural therapists may direct empowerment interventions to changing (or helping the client change) sociocultural and sociostructural systems or may direct interventions at the individual or relational level while focusing on sociostructural influences on the development of or meaning of problems.

The field of psychology overall has a relative emphasis on the individual level, such that most training, research, and therapy are conducted at this level. The dominance of an individual emphasis in the field reflects the individualism of the European American cultures within which psychology and psychotherapy were developed. While still true today, we are increasingly expanding our understanding of the influence of contexts on experiences and how our understanding itself reflects changing social norms and contexts. This is particularly so in relation to understanding how to approach individuals and groups who have less privilege and access to resources. For example, before the women's movement, White middle-class women who were not content to stay home and be housewives and mothers were frequently viewed as pathological. Such women might have been viewed as overly ambitious, lacking in healthy nurturance and caring, overly masculine, and poorly adjusted to being a woman. This "psychopathology" has been re-interpreted as having a sociocultural cause—that of sexism and constrained social roles and expectations for women. The intervention of the women's movement addressed the problem at the sociostructural level.

While we want to understand how contexts affect the development of the problems of an individual, these problems may be expressed or experienced in multiple ways, and there will always be varying opinions about how they began and developed. It is important to realize that there are always multiple factors—from individual, family, different contexts—contributing to the manifestation of problems.

We also need to consider who and how it is decided that something is a "problem." Does the client perceive the problem and seek help? Or is it a family member? Or is it a teacher or some other authority figure in the client's life? What has been done to try to resolve problem(s) prior to entering therapy? Sometimes, labeling something as a problem becomes the problem in itself, or reactions to problems or attempts to address them are actually the most critical issue. For example, when Mark's (age 15) mother found pot in his closet, she became enraged. She grounded him for a month and searched his clothes, bedroom, and closet diligently every day. She insisted that he enter therapy as she felt him to be troubled because he had smoked pot. The therapist found that the mother's heightened intrusiveness and constant haranguing became a more serious problem than the pot smoking, which was experimental and occasional. He empathized with the difficulties of raising a child as a single working parent and was able to help the mother see that work was required on the mother/son relationship, not just on the issue of Mark's pot smoking.

Your relative emphasis on the most important areas to attend to within contexts in conceptualizing clients will affect the choices you make for interventions. Each context affects clients' thoughts, feelings, and behaviors. In addition, each context has current influence and experience as well as past influence.

Focus of Change

Here we consider whether the focus of change is on behavior, cognition, or affect (emotion). A focus on behavior involves understanding and changing what people *do*. A focus on cognition involves understanding and changing what people *think*. A focus on affect involves understanding and changing what people *feel*. Most long-lasting change will affect all three realms (see Arnkoff, 1980), but most therapies tend to relatively emphasize one of these areas over the others. This emphasis may be reflected in what area is initially explored, or in a continued area of focus, in which case the therapist usually understands the other areas to be inherently influenced by change in the targeted area. Even those therapies that explicitly incorporate more than one foci often have a relative balance or prioritization.

A focus on behavior is usually most easily understood (although perhaps not as easily enacted) often because behavior is most visible from an outsider's point of view. A focus on cognition includes not only what a person thinks, but also how he or she thinks, for example, how a person attends to or does not attend to different things. In addition, a cognitive focus may primarily include the ways that thoughts are linked together into schemes or stories, as in narrative therapy, where the overall story and meaning made through connections is the primary focus rather than particular thoughts or thought patterns.

Affective change is frequently the most difficult for novice therapists to understand. Conceptualizing affect means focusing on the ways that a person feels about him or herself, relationships, and contexts. But it is difficult to conceptualize affect without describing it through observations of thoughts or behaviors. However, interventions focused explicitly on affect look different than those focused on thinking or behavior. Emotion focused therapy interventions aim to bring difficult feelings into awareness or to language, or facilitate the client to experience the feelings directly, frequently in the present moment of therapy (Greenberg, 2002). The more relational psychodynamic theories (for example, Object Relations) emphasize creating affective changes through the experience of the therapy relationship.

It is sometimes difficult for novice therapists to see "just being with someone" as a means for creating change. It may be more comfortable to focus on techniques and strategies—doing rather than being. However, it can be helpful to actively consider how a positive therapeutic relationship may make a person feel and then consider how these feelings may contribute to change more generally.

Obviously, your focus will relate to whether you are more comfortable exploring and changing thoughts, feelings, and/or behaviors in your own life. Your comfort level relates to your patterns of relationships, how your parents and teachers attempted to "teach" you, as well as to aspects of your temperament and personality style.

EXERCISE 3.16 ■ Choose a moment or event from the last few years of your life and share with a partner. What are you sharing? Consider with your partner whether you are describing thoughts, feelings, behaviors, or some mixture of these things. Is there a relative emphasis? As you consider the experiences of you and your partner, you may learn how different people emphasize different things. Now, ask each other to expand on each area of description: thinking, feeling, doing. What is most comfortable to ask about? What is most comfortable to describe?

EXERCISE 3.17 ■ Return to your list of Nancy's problems and the things you would like to know more about. Draw three columns on a sheet of paper and label them Affective, Cognitive, and Behavioral. See if you can categorize the lists you made in previous exercises into those dimensions; some will overlap, obviously, but what do you see as relative emphasis? Which of Nancy's problems do you think that you would be most comfortable addressing in which dimension? How does your comfort level resonate with what you see as important priorities for Nancy in therapy?

While you may have a preferred dimension, it is important to develop conceptualization and intervention skills in the other areas, in order to actively consider how these three dimensions interact and to recognize when and how you need to shift your attention. The area that you are most comfortable with will influence what you explore first with your clients and what is easiest for you to understand. It will also affect your adoption of different aspects of established theories and the things you actually do or say in therapy.

Consider the case of the married couple Eileen, age 53, and Tim, age 50. They came to therapy because Eileen had discovered that Tim was chatting online with other women and she presumed that he was engaging in online sex. Eileen was the oldest of six children from a working class Irish Catholic family and Tim was the only child of a divorced Jewish professional class family. Each had been married before, although neither had children from their previous marriages. Immediately after their wedding, they began years of infertility treatment in order to have a child. Now, they have a 7-year-old daughter. In the intake sessions, Eileen accused Tim of not being supportive, and of being self-centered and unfaithful. She acknowledged that her father and first husband were alcoholics, that she had been the caretaker throughout her childhood and adolescence for her younger siblings and that her mother had left the country for a year, leaving Eileen when she was 13. She felt she could no longer be empathic and understanding to a deceiving male. Tim admitted his chat room behavior, which he did not perceive as being unfaithful, and said that he and Eileen had not been emotionally close or intimate since their daughter was born. Tim also acknowledged his difficulty in expressing his feelings and being supportive. He described an emotionally abusive childhood with an overbearing, cold mother and a father who was emotionally unavailable in his early childhood and then physically absent following his parent's separation when he was 5. His parents were then involved in a high conflict custody case where neither seemed interested in Tim's feelings. Tim's father died when he was 10 before the custody case was resolved.

Tim and Eileen were asked to independently complete a list of the pros and cons of either remaining in or ending the marriage and a list of what each wanted the other to change and what each was willing to change. Eileen wanted to remain in the marriage because her religion forbade divorce and because she wanted to maintain the current lifestyle for her daughter. Tim wanted to remain in the marriage only if they could regain some of the intimacy and companionship that had originally attracted them to each other.

The therapist focused primarily on the family systems context. Within this, he addressed affective, cognitive, and behavioral aspects in conceptualization and treatment planning. Affectively, the therapist focused on how both Tim and Eileen maintained a distance from their own emotional experience; the ways in which they were both fearful of exposing emotional vulnerability in their relationship were connected to the increased difficulty of accessing or experiencing their own feelings. Cognitively, both Tim and Eileen had beliefs about their own behaviors as well as the other's behavior. Tim believed that Eileen was only interested in him for his money and Eileen believed that Tim was deliberately deceiving her and on the verge of abandoning her. Both had developed behavioral patterns of blaming and avoidance that the response of the other reinforced.

We present this case because frequently novice therapists believe that a relative emphasis on affect, cognition, or behavior determines one's theoretical orientation. While in some cases this may be true (for example, a behavioral orientation relates to an emphasis on behavior), there are many ways in which the focus of change can interact with but not determine one's theoretical orientation.

Relation of Focus of Change Dimension to Established Theories Therapy theories generally assume that change in one area (the area that a particular theory emphasizes) will ultimately create change in all three areas. The relative emphasis within an established theory is related more to how problems are initially conceptualized and how interventions are initially directed than to an idea of *constraining effects* to one area.

Psychodynamic theories are most consistent with an emphasis on cognitive and affective change, with interventions directed towards insight through interpretation and affective shifts through the therapeutic relationship. Traditional cognitive and cognitive-behavioral approaches (for example, Ellis' REBT (Rational Emotive Behavior Therapy) and Beck's Cognitive Behavioral Therapy (CBT)) are most consistent with a primary emphasis on cognition and a secondary emphasis on behavior for both conceptualizations and interventions. When interventions are directed at behavioral change, awareness of a concomitant change in cognitions may be seen as important for generalization of that change (consistent with an emphasis on the cognitive realm). Behavioral interventions are most consistent with conceptualizing problems as and directing interventions at behaviors; they vary in their secondary attention to conceptualizing cognitive or affective problems or changes associated with behaviors.

Humanistic and phenomenological therapies (including existential, experiential, and emotion-focused therapies) are consistent with a relative emphasis on affect. Perls' (1969, 1970) gestalt therapy is a good example of how an established theory might emphasize affect in conceptualization and intervention, as this approach makes explicit the importance of experiencing emotion as part of addressing "unfinished business" and includes explicitly affective oriented techniques (for example, the verbal command to "stay with the feeling"). Greenberg's (2002) process experiential therapy is a more recent example of the emotion focused therapies; they, too, explicitly emphasize experiencing affect in the moment. Process experiential therapy is also an example of a constructivist therapy.

Systems theory, feminist theory, constructivist therapy, and multicultural theories are consistent with emphasis on all three dimensions: behavior, cognition, or affect. The central tenets of these theories do not necessarily emphasize a particular pathway for change, as much as place emphasis on where and how a problem develops and is maintained (for example, relationally, sociostructurally), and by epistemological and philosophical stances on human development and change. Specific theories within these umbrella approaches may be more or less consistent with emphasis on cognition, affect, or behavior for initial understanding of the problem or intervention (for example, family narrative therapy, cognitive constructivist therapy, process experiential therapy).

EXERCISE 3.18 ■ Without looking at your lists, try to remember what your impressions of Nancy's issues were and are. Does this tell you anything about your emphasis on thought, behavior, or affect?

Past or Present Emphasis Now, we will address the balance of emphasis on the past or the present. Some therapists believe that understanding how a behavior, thought, or feeling developed from past experiences is centrally important to creating change. They believe that understanding the influence of family context or sociocultural/sociostructural context developmentally (or historically) is as important as understanding the current contextual influences. Others believe that understanding the past is not necessarily important; what is most important is understanding the current experience and function of the clients' behaviors, thoughts, or feelings. Like all of the issues described here, this is not a dichotomous issue, but one of relative emphasis, and is better conceptualized as a continuum.

EXERCISE 3.19 ■ Think about your favorite novel. What is the focus? What is the structure? Do you learn about the main character's past, or is it mostly focused on the current events? What makes this your favorite novel, and how does this relate to your interest in the past or the present?

EXERCISE 3.20 ■ Return to Exercises 2.3 and 2.4. In describing your experience and approach to change, did you describe an understanding of what past experiences, relationships, or circumstances contributed to the development of the problem? In your process of change, was it important for you to understand the developmental contributors and influences to the problem in order to decide how you would address it?

Therapists who emphasize the role of the past frequently see current thoughts, feelings, and behaviors as strongly embedded within a developmental network of associations. They frequently value the role of conscious insight into this development, believing that current change is easier or made possible if one understands how and why the problem developed. They may, therefore, devote time in therapy to exploring the past and helping clients to develop this conscious insight.

Therapists who more strongly emphasize the present do not discount the past as important influence and often agree that conscious insight into problem development can be helpful. However, they do not believe that this insight is necessary for change and sometimes believe that it is distracting or defensive. Therefore, they focus more on the current experience of the client, changing current thoughts, feelings, or behaviors.

EXERCISE 3.21 ▪ Return to Exercise 3.13. What kinds of things were you most interested in knowing more about in order to understand Thanh? Were your questions focused on the past or on the present? Because you were asking initial questions with very limited information, you might be more likely to focus on understanding the current "presenting problem." But consider how you would work with Thanh over several sessions. Would you feel strongly that you would want to understand his childhood and developmental relationships and experiences? Would you focus only on those relationships and experiences that directly relate to his current problems, or would you feel strongly that you wanted a broader understanding of his development? Consider interactions of your preference for past–present focus with your understanding of contexts. For example, would you want to know about his family's past or focus more on current family influences?

In the case of Tim and Eileen, the therapist felt it was important to connect the present to the past so that each could gain some compassionate understanding of the other's experience in developing views of self, self-in-relation, and of the world. Tim, for example, learned at a very young age to bury his feelings in order to endure his mother's rage; his present relationship shows that this pattern no longer works and that the wall is not just between him and others, but also between himself and his feelings. Eileen's traumatic experiences having to manage a household at the age of 13 with an abusive alcoholic father and a mother who abandoned the family affects her current relations in that she does not trust anyone and needs to be in control in order to maintain the financial security and social status she lacked.

Relation of Past or Present Emphasis Dimension to Established Theories No theoretical orientation is completely oriented to the past, as the client's current experience and presenting problem, effects on daily living and functioning, and the therapy itself are happening in the present. So to a certain extent, the question is, how much do different theoretical orientations (or therapists) include the past either as a necessary area for understanding the problem or as a major influence on the type of intervention?

Psychodynamic theories are most consistent with a relative emphasis on the past. Rationalist cognitive, cognitive-behavioral approaches, and behavioral theories are largely unconcerned with the past as explicit material for conceptualization or intervention. It is not necessary to know how a problematic thought or behavior developed. Cognitive constructivists frequently fall more along the middle of this dimension, with greater attention to conceptualizing the developmental context, both for the therapist and for the client. Similarly, phenomenological systems feminist and multicultural theories are consistent with emphasis on both past and present. Specific therapies within these groups, or particular therapists within theoretical orientations, may emphasize the past more than others.

EXERCISE 3.22 ■ Now, go review your lists of problems for Nancy and for Thanh. How much of their pasts would you want to explore, how, and why? Look at your lists and discuss within small groups what your thinking is about the location of the problem and the pros and cons of exploring the past. How might your thinking be different for each of them? What factors might contribute to your different thinking?

We want to emphasize that while you may be more comfortable emphasizing certain dimensions, each case is different and the client's needs may require you to focus on whatever dimension is most helpful to the client. For example, while you may emphasize the present, a particular client may feel that it is most helpful to make connections to past experiences that help him or her to understand the development of current patterns of behavior which, in turn, opens up possibilities for modifying the interaction of current thoughts, feelings, and behaviors.

SUMMARY

In this chapter, we have explored an ecological model and the multiple ways that contexts might affect experience. We particularly explored the family and sociocultural/sociostructural system contexts and considered how attention to different levels of context including power and privilege might affect our conceptualization of clients and their presenting problems. In applying understanding of contexts, we utilized the case of Nancy as well as additional cases of Thanh, and Tim and Eileen. We discussed the conceptualization of the location of the problem in relation to different contexts, specifically the individual, family, and sociocultural/sociostructural contexts. We then explored focus within contexts, considering differing emphasis on affect, thought, or behavior and on the past/present continuum. Throughout the chapter we emphasized that therapists need to be flexible in their conceptualization and intervention plans in order to focus on what will be most effective for each client in his or her unique contexts.

REFERENCES

Arnkoff, D.B. (1980). Psychotherapy from the perspective of cognitive theory. In M.J. Mahoney (Ed.), *Psychotherapy processes* (pp. 339-361). New York, NY: Plenum.

Bronfenbrenner, U. (1979). *The ecology of human development.* Cambridge, MA: Harvard University Press.

Crawford, M. & Unger, R. (2004). *Women and gender: A feminist psychology* (4th ed.). New York: McGraw Hill.

Constantine, M.G. & Sue, D.W. (Eds). (2006). *Addressing Racism: Facilitating Cultural Competence in Mental Health and Educational Settings.* Hoboken, NJ: Wiley & Sons.

Crenshaw, K., Gotanda, N., Peler, G., & Thomas, K. (1989). *Critical race theory: The key writings that formed the movement*. New York, NY: New Press.

Greenberg, L. (2002). *Emotion-focused therapy: Coaching clients to work through feelings*. Washington, DC: American Psychological Association Press.

Knopf, H. (1986). Identifying and classifying children and adolescents referred for personality assessment. In H. Knoff (Ed.), *The assessment of child and adolescent personality* (pp. 1-28). New York: Guilford Press.

McGoldrick, M., Giordano, J. & Garcia-Preto, N. (Eds.) (2005). Ethnicity and family therapy (3rd ed.). New York: Guilford.

Nichols, M.P. & Schwartz, R.C. (2006). Recent developments in family therapy: Integrative models. In *Family therapy: concepts and methods*. 7th ed. Boston: Pearson/Allyn & Bacon.

Okun, B., Fried, J., & Okun, M. (1999). *Understanding Diversity: A learning-as-practice primer*. Pacific Grove, CA: Brooks/Cole.

Okun, B.F. & Kantrowitz, R. (2007). *Effective helping: Interviewing and counseling techniques*, (7th ed.). Pacific Grove, CA: Brooks/Cole.

Root, M.P.P. (1998). Reconstructing race, rethinking ethnicity. In A. S. Bellack & M. Hersen (Eds.), *Comprehensive clinical psychology* (pp. 141-160). New York, NY: Pergamon.

Schervish, P. G. (2011). The Family Wealth Impact Project. Report of the Boston College Social Welfare Research Institute.

Sue, D.W. & Sue, D. (2007). *Counseling the culturally diverse: Theory and practice* (5th ed.). New York, NY: Wiley.

Tannen, D. (1996). *Gender and discourse*. New York, NY: Oxford University Press.

Uba, L. (1994). *Asian Americans: Personality, patterns, identity*. New York, NY: Guilford.

4

Conceptualizing Therapeutic Relationships

Conceptualizing involves not only understanding yourself as a therapist and the particular experiences of the client in context, but also your own relational style as a therapist and the ways in which you understand the role of the relationship in contributing to therapeutic change. Therapists from diverse theoretical orientations recognize the importance of the relationship and the therapist's conceptualization of the role of the relationship in contributing to change in psychotherapy. In fact, the research is quite clear that the therapeutic relationship is a significant context for any approach or intervention (Gelso & Hayes, 1998; Mallinckrodt, 2010; Norcross, 2002). As in all of the dimensions of conceptualization, it is important to be aware of your own values and feelings about relationships, shaped by your own culture and experiences. Your self-awareness will influence how you view and understand your clients.

EXERCISE 4.1 ▪ What, to you, makes a good relationship? Describe the elements you feel characterize a good relationship in the following roles: parent/child; intimate partners; mentor/mentee; friends; supervisor/supervisee. What elements do you think would contribute to a bad or harmful relationship? Are there similarities in what you think is valuable across different types of relationships? What are the differences in the things you feel characterize good relationships in these different roles? Now, make a list of what you think are the characteristics of a good relationship between therapist and client. Compare your list with at least two of your peers.

UNDERSTANDING YOUR RELATIONAL STYLE

Everyone has different styles of relating, for example, some people are extroverted and others are more introverted, some talk more and others talk less. Your relational style as a therapist may be different in some ways from your general style because of the specific role. For example, people frequently talk and disclose less in their role as therapists than they might in personal relationships. However, your therapeutic style will be affected by your own preferences. Your orientation will also affect your style and vice versa—your relational style will influence your beliefs about change and how the relationship contributes to change.

In the first part of this chapter, we will focus on four dimensions of interpersonal style related to how you interact with clients as a therapist. These dimensions include the extent to which you are: 1) directive in content; 2) structured in process; 3) active in sharing your thoughts, understandings, or feelings with clients; and 4) confrontative. Your preferences within these dimensions and your thoughts about what preferences will best contribute to change will affect how you will act in the therapy room with a client.

We recognize that most therapists have a range within each of these dimensions in which they are comfortable, rather than a single point or experience. Therapists tend to move within that range according to the client's needs and style, as well as according to the development of therapy over time. For example, many therapists are less directive, structured, and active in the beginning of a course of therapy, as they want to gather as much information as possible and understand what is most important to the client. However, they may become more directive, structured, or active as they develop a fuller conceptualization of the client in context and a plan for intervention and change. Our experience is that being on either extreme end of any of these dimensions is likely to be less effective for most therapists with most clients.

As you develop your own self-awareness on these dimensions, consider also how you come to understand your client's preferences on these dimensions and their responses to you. If you ask clients at the end of their first session how they experienced the session and what their expectations had been, you might get responses such as "I thought you would tell me what to do," or "I liked that you talked to me about what I was saying," or "I'm glad you didn't push me."

Directive versus Non-directive

This dimension addresses the extent to which a therapist directs the *content* of interactions. One way to describe this is the extent to which the therapist determines *what* will be discussed or addressed, as differentiated from *how* it will be discussed or addressed (see *Structuring Process, next section*). All therapists influence the client's choices about what to discuss to some degree, even if they are not intentionally doing so. For example, if a therapist nods and leans forward when a client is discussing his or her relationship with his or her parents and does not nod or lean forward when a client is discussing schoolwork, the therapist is (perhaps unintentionally) communicating to the client the belief that talking about the relationship with parents is a more fruitful direction than talking about schoolwork.

The dimension of directive versus non-directive, however, is less concerned with what direction you might encourage, and more concerned with the extent to which you set or encourage a direction. More directive therapists will be more active in encouraging clients to address certain types of content while less directive therapists will be more likely to allow the client to set his or her own direction and determine what is most important to him or her. The actual content that the therapist is encouraging will depend on other aspects of the therapist's beliefs about change. For example, a therapist who emphasizes cognition will direct the client to discuss his or her ways of thinking, while a therapist who emphasizes the role of the past will direct the client to discuss developmental experiences and relationships.

EXERCISE 4.2 ▪ Rate yourself on the following statements with 1 being "I strongly disagree" and 5 being "I strongly agree":

1. I am comfortable with ambiguity and uncertainty.
2. I believe I am an organized thinker.
3. I gather information and then organize.
4. I organize a framework and then gather information that is related to that framework.
5. I am a concrete thinker.
6. I am an abstract thinker.
7. I follow my intuition comfortably.
8. I wait for evidence before making decisions.
9. I am patient waiting for others to "get it."
10. I prefer to see results quickly.
11. I am uncomfortable if I do not quickly understand how things relate to one another.
12. I like to explore new ways of getting from one place to another, even if I sometimes get lost.

Which of these items do you think pertain to the directive end of the continuum and which to the non-directive end of the continuum? How you answered these questions may help you see where you fit on the directive/non-directive continuum. If you found yourself wanting to answer "sometimes" or "in some ways," consider when or how, and what this might tell you about how you will approach therapy with different clients. It is helpful for you to think about these questions not only as a helper, but also as a trainee with regard to supervision and teaching.

Wherever you fit on the directive/non-directive continuum, as long as you can adapt your stance to the client's needs within the treatment context, you are likely to be effective. With such adaptations, you may come to find that you are comfortable and competent in a wide range of therapeutic stances of direction. In training, we often suggest to trainees more comfortable with a non-directive approach that they practice more directive approaches and vice versa. This is helpful particularly because we are usually most comfortable with what we are most accustomed to, but may find that we can be comfortable with a wider range once we have experience with different approaches. As in other dimensions, we believe that being at either extreme can be less beneficial to clients. For example, if one is always completely non-directive, the client cannot benefit from the expertise and training of the therapist. And the therapist may inadvertently enable the client to avoid self-responsibility and contribute to barriers to change. Likewise, if one is completely directive, the therapy cannot be tailored to the client's particular person, context, and presenting issues; any changes may not be internalized in order to persist.

Structuring Process

This dimension addresses the extent to which the therapist structures *how* material is engaged or addressed. It is related to the directive/non-directive dimension because both dimensions are concerned with guiding clients' attention. But structuring process is not about directing attention to consider particular content, but about directing different ways of exploring the content, whether the client or the therapist chooses the content. Traditionally, the process of psychotherapy has an inherent overall structure as a timed dialogue with a particular purpose. But there can be considerable variability in structuring within this overall framework.

Frequently, a less structured process approach means that the therapy consists primarily or solely of open dialogue. A more structured process is likely to involve the use of specific strategies to structure the exploratory or intervention process, such as role playing, story telling, metaphors, and imagery. Creative therapies such as art, movement, play, or music therapy are structuring the process of exploration through their particular modality. What is explored may or may not be guided by the therapist depending on the level of content direction (the dimension discussed earlier). What is important at this point is for you to be aware of your interpersonal style and comfort level with structured process as well as understanding of the client circumstances and contexts for which a more or less structured process would be beneficial, regardless of your preferences.

Another aspect of process structure is whether it occurs solely in the therapy room, outside the therapy room, both, or neither. Using inter-session assignments (homework)—such as keeping a journal, reading, trying out a new activity—reflects a more structured process approach. The thinking behind these activities is that change needs to take place in clients' lives more generally, not only in the therapy hour; learning requires continuous, multimodal activities. For example, when working with a client who is struggling with depression and having difficulty getting out of bed, you may suggest that he or she get out of bed in the morning, shower, dress, and so on and then if they want to go back to bed, they can. They only have to contract for the above specific behaviors. The thinking behind this is twofold: 1) activity is an antidote to depression; and 2) it is unlikely that someone will disrobe and immediately return to bed after having been successful at getting out of bed, showering, and dressing.

EXERCISE 4.3 ■ Think about your own comfort levels and rank the following with 1 being "I strongly disagree" and 5 being "I strongly agree":

1. I like to know what I am doing from one moment to the next.
2. I become bored if I do not have something specific to do.
3. I hate cooking with recipes and prefer to "just wing it."
4. I like to figure things out for myself.
5. I like someone to give me instructions on how to do what is needed.
6. Having unscheduled free time is my greatest pleasure.
7. I was always able to get my work done without reminders or supervision.
8. I don't like "going by the book."
9. I like to plan what I will do with my free time.
10. I like to cook with recipes.
11. I need things to be organized around me.
12. I like classes where the syllabus is detailed and the professor adheres to it.

How you answered these questions may help you see where you fit on the structured/unstructured continuum. Consider how your own preferences may affect your approach as a therapist. Do you think you will be different in your role as a therapist?

Level of Activeness

This dimension addresses how active the therapist is in a general sense, differentiated from direction or structure. Some therapists speak rarely while others interact more frequently. Some therapists are quiet people and prefer to listen and internally reflect, while others are more talkative and interactive in a preponderance of their relationships. Some therapists engage in activities with clients—that is, helping the client write a resume, teaching interview skills, encouraging the client to actively seek entitlement resources, and so forth. This dimension is related to the other dimensions, in that therapists who are more directive and structured will inevitably be more active. However, non-directive and non-structured therapists may be quite active as well.

EXERCISE 4.4 ■ Consider your own interactions with friends, with family, at work, in school. In what contexts are you more interpersonally active? In what contexts are you less active? With what degree of interaction are you comfortable? Do you ever find yourself impatient with others' rate of responsiveness or the amount of "air time" they take? Ask two people who know you well how they rate you on this dimension.

Although activeness as a therapist may relate to your style with your friends, colleagues, and acquaintances, a therapeutic relationship is a special kind of relationship and so your style may be different in this role than in other parts of your life. The role of therapist and the goal to help the client means that even therapists who are really outgoing "talkers" in their personal lives are careful and attentive listeners who talk much less as therapists than they do in other roles and settings. But therapists do vary in how active they are in their interactions with clients, and it is important to be aware of your tendency and comfort level.

Confrontativeness

Some people are very comfortable with confrontation while others actively avoid it. Comfort with confrontation means being comfortable with conflict at least to some extent, because confrontation is about engaging a difference of viewpoint. This difference may be in the conscious awareness of both people (active disagreement) or may be related to a lack of awareness, such as when someone "confronts" you with an interpretation of your behavior and you may not have been consciously aware you were even behaving that way.

EXERCISE 4.5 ■ Here, we are encouraging you to consider how comfortable you are with confrontation generally. Think about your own comfort levels and rank the following with 1 being "I strongly disagree" and 5 being "I strongly agree":

1. I like to debate ideas; vigorous debate and exploring disagreement is exciting.

2. While I appreciate when other people call attention to things that I am doing that I am unaware of, my first response is frequently feeling angry or hurt and wanting to withdraw.

3. In my close relationships, I have frequently had intense disagreements that are then resolved in ways that bring us closer together.

4. I am most comfortable when everyone agrees and has similar perspectives.

5. I appreciate when other people call attention to things that I am doing that I am unaware of; I respond to this quickly by wanting to know more about their perspective.

6. I think the best way to collaborate is to build on and expand each other's ideas.

7. I rarely have intense disagreements in my close relationships and remember very few times of raised voices or strongly charged emotion.

8. I think that many people arguing for their own perspective or solution leads to a richer exploration and is frequently the way to find the truly best solution.

9. I am suspicious when everyone seems to agree and wonder if someone is disengaging or hiding their true feelings.

Your comfort level with confrontation may also vary in relation to who you are interacting with (colleagues, strangers, family), the context in which you are in (for example, debate club is a very particular context), and whether you are discussing abstract ideas or more personal beliefs, behaviors, or feelings. In therapy, moments of confrontation are typically directive and active because, at these moments, the therapist is directing the client to consider something that is at least to some degree outside the client's current understanding. This does not mean that you cannot be directive or active without being confrontative. As we mentioned above, most therapists are infrequently directly confrontative, so we are really encouraging you to consider where you fall in avoiding confrontation.

Frequent or intense confrontation is usually uncomfortable for most people, particularly if they are on the receiving end of the confrontation. Because of this, most therapists are not highly confrontative, but there are certainly exceptions to this. Both Fritz Perls and Albert Ellis were frequently highly confrontative and both are also seen as master therapists. These master therapists used confrontation as a major strategy for creating change generally. But even therapists who are not as comfortable with confrontation and do not integrate it into their orientation or approach to change as a major strategy are sometimes confrontative.

When clients seem "stuck" or resistant to change, confrontation may seem like the only option. But it may be possible to be "confrontative" while working to avoid the more negative aspects of confrontation, when clients or therapists may feel a threat to the working alliance. Motivational interviewing (Miller & Rollnick, 2002) is a client-centered directive interdisciplinary method for helping people explore and work through ambivalence about changing. The model is based on Prochaska's change model (Prochaska, Norcross, & DiClemente, 1994) which presents a framework for stages of the change process: a) pre-contemplation, when the client is not yet considering change; b) contemplation, when the client is considering change, such as stopping smoking or going to the doctor, but is quite ambivalent; c) determination, when the client makes a commitment to change; d) action, when there is active involvement to change; e) maintenance of behavioral change, when the client works to solidify changes; f) relapse, when the undesired behaviors return; and f) termination, when the change is stable. There is no timetable for how long these stages may last. Motivational interviewing builds upon the understanding that clients may present with different readiness for change.

Motivational interviewing can give us a different understanding of confrontation, as it aims to "confront" a client's ambivalence about change that may be related to difficulties in taking steps that would contribute to change. However, motivational interviewing actually aims to be non-confrontative. This oxymoron is engaged by using what may be seen as confrontation in other approaches in a slightly different manner, that is, by working *with* clients to develop and explore discrepancy to help clients engage their ambivalence and develop their desire (motivation). For example, a therapist using motivational interviewing may ask questions such as: "What do you think about your drinking?" "What do you know about the risks of drinking?" "How do you think your drinking impacts the others in your family?" "What are the pros and cons of cutting down or quitting your drinking?" "On a scale of 1–10, how important is it to you to change?" Simultaneously, the therapist will make affirming statements such as: "I see how important this is to you….," "You really are trying hard to change," "It really is courageous for you to take this step," "So, let's talk about what we've covered today," "How have you dealt with problems in the past?" Thus, motivational interviewing may be one approach to helping clients become aware of resistance while also being less confrontative.

EXERCISE 4.6 ▪ Consider what the purpose of confrontation might be and when you might want to be confrontative. What might be some of the benefits or drawbacks to being confrontative with a client? Motivational interviewing aims to work towards the benefits while minimizing the drawbacks. Are there other ways that you interact with people in your life, or that you might interact with clients, to address some of the goals that confrontation might address while avoiding possible drawbacks?

Relation of Dimensions of Relational Style to Established Theories

The variability on relational style dimensions among therapists within a given orientation can be quite high, as many of these dimensions are not related to specific core concepts within established orientations, but are instead indirectly related (see Table 4.1). Furthermore, the position on these dimensions within an established orientation may be highly related to the developmental stage within therapy. Most theories are non-directive, particularly in initial stages of therapy, in that the client establishes the content. Both CBT and liberation theories are more directive because they direct clients to address particular *kinds* of content (cognitions or behaviors, or content related to sociostructural contexts), but they still do not direct the actual topics to be discussed. But at various moments within therapy, therapists within almost all orientations may be more directive: for example, a psychodynamic therapist may direct the

TABLE 4.1 Relation of Dimensions of Relational Style to Established Theories

	Directive vs. Non-directive	Structured vs. Unstructured	Activity Level	Confrontativeness
Psychodynamic	Less directive	Semi-structured	Lower	Moderate
CBT	More directive	Structured	Higher	Moderate to higher
Humanistic Client-Centered	Less directive	Unstructured	Lower	Lower
Humanistic—Existential	Varies	Structured	Varies	Moderate to higher
Systems—Ecological	Varies	Structured	Moderate	Varies
Constructivist	More directive	Varies	Moderate	Varies
Liberation Theories	Somewhat directive	Varies	Moderate	Moderate

content of a given session to a topic that had been identified as central, or one that has been avoided.

Most established theories have some structure to them beyond the general framework of a dialogue, but there is great variability even within the same theoretical orientation. The activity level for established theories, as noted above, relates strongly to the directive and structured dimensions. Confrontativeness is lowest in humanistic client-centered therapy, related to the core idea that clients have a tendency towards self-actualization, thereby reducing the need for therapist confrontation of resistance to change or lack of awareness. Other established theories vary in confrontativeness, and vary even more in how confrontativeness is enacted. For example, interpretations are a kind of confrontation, as they bring into awareness issues that clients are frequently not aware of.

Relational Style Dimensions and the Role of the Therapist

These dimensions of therapists' relational style are interdependent and are affected by how you see the role of the therapist and how you understand the development of psychological problems and your own theories of change. Conceptualizing yourself as a therapist overall means considering how you see your role in relation to the client, including understanding and comfort with the power that is inherent in the role of therapist. Therapists do have great influence on clients, and the role of therapist gives them credibility in determining what is healthy or good for the client. This power influences clients and therapy whether or not you are directive, structured, active, or confrontative in your stance.

It is also important to be aware of the power that we have to potentially enable clients to create or maintain patterns of thinking, feeling, or behaving that are harmful to them. Some people come to therapy because they want validation in their blaming of others for their problems and do not want to assume any responsibility for their own role in creating a solution. How do we help these and all clients understand that they cannot force others to act differently, although perhaps changing their own actions will elicit change in others? How can we help clients feel empowered and motivated to take responsibility for what can be changed? As we delve into the therapeutic relationship, it is important to understand that while we are developing continuously an empathic, supportive working alliance, we need to be sure that we do not avoid or minimize difficult issues or happenings in therapy because we do not want to "upset" the client and possibly have him or her not return. There are many reports of clients who remain in therapy for a long time without the therapist finding a way to help the client take responsibility for his or her own part in whatever the problems are or for creating positive change. Our point is that some of the elements of other types of relationships—politeness and avoidance of challenging topics—have different meanings within the therapeutic relationship.

Some therapists understand the therapist to be an expert or a teacher who has extensive knowledge to share with the client in order to benefit the client. These therapists are more likely to be directive and may be more active, structured, and confrontative as well. Other therapists understand the role of the therapist to be more of a facilitator or a guide, seeing the client as being more knowledgeable about him or her self than the therapist and able to find his or her own answers with guidance.

EXERCISE 4.7 ▪ When you conceptualize yourself as a therapist in relation to clients, which of the following roles fits best?

Guide	Teacher	Facilitator
Friend	Goad	Manager
Gardener	Parent	Director
Collaborator	Student	Other: (describe)

Write a few sentences describing how this role applies to you. For example, "For me, being a therapist is like being a guide because…." How are guides and therapists similar? Then, consider what this choice tells you about your preferred style, level of comfort, and understanding of psychotherapy and yourself as a therapist.

UNDERSTANDING THE ROLE OF THE RELATIONSHIP IN CONTRIBUTING TO CHANGE

A therapeutic relationship is a particular type of relationship, as the primary goal of the relationship is to help the client. There has been considerable theorizing and related research exploring the therapeutic relationship that has attempted to describe aspects of the relationship that both cut across different theoretical approaches and can be used to differentiate preferences for theoretical approaches. These aspects include: the working alliance, the significance of the therapeutic relationship, the real-unreal relationship, and the process emphasis.

The Working Alliance

One of the most important aspects of the therapeutic relationship is the working alliance. A working alliance is "an authentic, warm, empathic relationship between the helper and the helpee" (Okun & Kantrowitz, 2007, p. 4). A working alliance reflects shared goals and a trust between the therapist and the client that both are working towards those goals. It is part of the real relationship (see below) between client and therapist. It is likely that at least some of the characteristics of a good therapy relationship that you described in Exercise 4.1 are related to the basic aspect of a good working alliance: trust, warmth, empathy, genuineness, and a feeling of working together.

How we shape a working alliance in therapy is related to our values about relationships and our comfort in being in different roles in relationships.

EXERCISE 4.8 ▪ Answer the following ranking them 1 for "I strongly disagree" to 5 for "I strongly agree":

1. I like competitive situations.
2. I like being given clear directions of what to do.
3. I like to be around people who are more competent than I am and know more.
4. I like others to need me.
5. I prefer working alone rather than on a team because I know the job will get done.
6. In team situations, I prefer others to take the lead.
7. I am an initiator more than a follower.
8. I am comfortable in a position of authority where I can tell others what to do.
9. I am very sensitive to criticism.
10. I like to please people so they will like me.

What do your answers to these questions tell you about your attitudes towards authority, hierarchy, individualism, or collaboration? How have your life experiences shaped these attitudes? How would they be different in personal, social, public, and professional relationships? Furthermore, given your attitudes, what are your expectations of clients?

EXERCISE 4.9 ▪ How do you think you could develop an empathic, effective working alliance with Nancy? What relationship skills and strategies will you need?

What might hamper this relationship? How does this fit into your theoretical perspective?

Relational norms, expectations, and values can vary significantly due to different cultural and background experiences. There may be pitfalls when you are working with clients whose identities and cultural/status backgrounds are significantly different from your own, particularly if these differences affect your clients' values and expectations about relationships.

EXERCISE 4.10 ■ In small groups, consider different cultural perspectives on the following relationship aspects:

1. the amount of intimacy encouraged in different kinds of relationships
2. how closeness is expressed or not expressed
3. the amount of expressed conflict or confrontation
4. direct and indirect forms of verbal communication
5. the management of facial expressions and nonverbal behavior
6. the amount of self-disclosure

EXERCISE 4.11 ■ Now, consider how these aspects and possible differences might affect four types of relationships: 1) social relationships with friends; 2) relationships with those you are supervising and need to discipline; 3) relationships with your boss or supervisor; and 4) relationships between a client and you as a therapist.

Sometimes cultural or personal differences in expectations can completely stymie the development of a working alliance. For example, if you place a high value on egalitarian relationships while your client prefers hierarchical relationships, you may have difficulty establishing the trust and sense of common goals that are necessary for a good working alliance. Val came to therapy to decide whether or not to leave her husband. She had been raised in a southern, rural blue-collar community; her family belonged to a fundamentalist religious group. Val had married a member of that group and followed him to a northern military base where he was stationed. Val wanted the therapist to tell her what to do; she spent six sessions detailing her complaints and dissatisfactions and became very angry when the therapist tried to get her to engage in processing the pros and cons of her situation. The therapist was a male, a military officer, and "he should know what the right answer was," repeated Val again and again. The therapist was unable to establish a working alliance with Val and suggested she might work better with a female therapist. Val, who struggled with her cultural beliefs that "therapy is bad" refused the suggested referral. She came from a culture where the rules and roles were strictly prescribed and she wanted a male authority figure to make a decision for her, but the therapist had been socialized professionally to collaborate with Val on problem solving. The two were therefore at a relational impasse that did not have much chance of resolving, particularly given the treatment context of eight sessions. These kinds of differences can also cause challenges in agreeing upon the goals of therapy.

Developing and maintaining a working alliance require different capacities. As the relationship develops, the client hopefully will feel more comfortable and may exhibit some behaviors or reveal some things that cause you to feel anxious, judgmental, or challenged. How will you deal with this? You may have to work extra hard to join without judging and probe the client's meanings and intentions. Thus, you can never assume the relationship is constant. One way to maintain an effective relationship is to "check in," seeking feedback, and ask something like "How are we doing?" "How are you feeling about me or the work we do here?" Another way is to monitor how open you are to verbal and nonverbal signals from the client about their feelings and thoughts. The cues we receive from clients—implicit or explicit—provide invaluable feedback about what we are pursuing, our pace, and our strategies. Like any interpersonal relationship, being attuned to the other person's reactions and responses enables us to modify and adapt our behaviors.

The working alliance is the foundation of all helping relationships, regardless of theoretical orientation. Thus, it is not a dimensional preference as are the other aspects of relationships discussed in this chapter. Learning to establish an initial positive relationship and working alliance is usually the first step in training for professional helpers. (See Okun & Kantrowitz (2007) for a thorough approach to developing skills for a working alliance.) But professional psychotherapists also need to conceptualize their therapeutic orientation and interventions in relation to other dimensions of the therapeutic relationship and their contributions to change such as: 1) the significance of the therapeutic relationship; 2) the real/unreal relationship; and 3) the process emphasis. The first two of these dimensions are taken from Gelso and Carter's (1985, 1994) excellent reflections on the therapy relationship. The third dimension draws heavily from Yalom's (2005) understanding of relational process. All of these dimensions are concerned with the therapist's understanding of the role of the therapeutic relationship in *creating change*. These dimensions relate also to one's understanding of how problems develop and how relationships in general (not just therapy relationships) affect individuals.

Significance of the Therapeutic Relationship

This dimension addresses the therapist's conceptualization of the significance of the relationship for the creation of therapeutic change. Is a particular type or quality of relationship *necessary* for change? Is it *sufficient* to create change? The first question addresses what Gelso and Carter call *centrality*—how important is the relationship (and a particular type of relationship) to the process of change.

The second question addresses what Gelso and Carter call the "means-end" dimension, which addresses whether the relationship is seen as the means for other mechanisms of change, or as the primary change mechanism itself.

EXERCISE 4.12 ▪ Share with a partner an incident in your life where a significant relationship brought about or influenced change in you. You might think about a parent, a teacher, a friend, lover, supervisor, boss, or mentor. How and why was this particular person and relationship influential? What elements of the relationship were significant? How? After you have completed this, share an incident where you feel you have contributed to someone else's change. What are you aware of about these relationships?

To see the relationship as not necessary means that a particular type or quality of relationship between the therapist and the client is not necessary beyond the working alliance. It means that change happens because of other things, such as shifts in ways of thinking, exploration of family roles, trying new behaviors, or learning about oppressive systems.

EXERCISE 4.13 ▪ There are some situations where the client's strong motivation and determination may compel him or her to try whatever tools are offered to achieve a goal. Share with a partner a time when a significant change occurred in your life outside of the context of a particularly good relationship, where the basic working alliance between you and a coach, teacher, or other role model enabled you to learn and change even if this relationship was unimportant or even if you did not even like the other person involved. For example, many sports players talk about their determination to succeed despite a parent, or coach, or someone whom they considered to be an impediment in some way. Think about the elements of motivation and determination and how they were fostered by lack of good relationships or by negative relationships.

While it is certainly possible to change in spite of difficult relationships, in many therapies, aspects of the therapy beyond the working alliance are a contributor to change. Maria came to therapy struggling with an episode of intense depression. She had had similar episodes of depression in the fall, at the start of every school year during her college years, but had not had an episode of depression in the previous fall when she had taken a year off after college. Maria had recently started graduate school and she believed that she had an obligation to her parents and to her cultural community to obtain her doctorate, because her immigrant parents had sacrificed so much for her schooling and because there were so few Latinas who had the opportunity to obtain a Ph.D. In exploring these beliefs and her relationship with her family, it became clear that Maria believed that, particularly as a Latina, her own needs must be sacrificed for the sake of others. Maria described her mother as "like a slave" to her husband and family. Her mother had communicated to Maria the message that husbands and fathers, even if abusive, should be indulged and accommodated, even if it meant denying or suppressing one's own emotions. For example, when Maria was a child and there was conflict between her parents, her father would want most to be with her and hear her say how much she loved him. Although she made it clear to her mother that these moments were hurtful to her and she frequently hid to avoid them, her mother would bring Maria to her father and encourage her to say she loved him regardless of what she was feeling at the moment (frequently fear, anger, hatred) in order to avoid angering her father.

The relationship with the therapist, who was also a woman of color with a Ph.D, was central to Maria's change in several ways. While the therapist honored Maria's cultural values and connections, she encouraged exploration of Maria's belief that there was only one way to respect and give to her family and community. The therapist's own willingness to explore and accept the importance of Maria's culture, her connection to her family and her ambivalence about pursuing a Ph.D., and ultimately to validate and support Maria's choice to pursue a master's degree in a different field enabled Maria to imagine different possibilities.

The centrality of the relationship was even more evident in Maria's change in her belief that close relationships meant sacrificing herself to others. While exploring her feelings and thoughts about her family and partner relationships was important, a turning point occurred when Maria's schedule changed and she could no longer come to therapy during regular business hours. Rather than discussing this with the therapist, Maria started a session talking in a distant manner about ending therapy. As this was quite "out of the blue," the therapist focused on exploring Maria's thoughts and feelings leading to this announcement. Maria eventually tearfully confessed that her schedule had changed and, while she did not want to end therapy and was fearful of relapsing if she did, she assumed that the therapist would not be able to accommodate an "after hours" appointment. The fact that the therapist gladly did accommodate this need and encouraged Maria to directly express her needs in therapy challenged Maria's beliefs about relationships and thus contributed to therapeutic change. This case is a prime example of how the therapeutic relationship can be the foundation of therapeutic change.

EXERCISE 4.14 ▪ Returning to Nancy's case, jot down what might be interpersonal aspects of your relationship with her, from both your perspective and hers. What do you think the significance of the therapy relationship is for her? Discuss your thoughts and feelings about this significance with others and note differing perspectives.

Relation of the Significance of the Therapeutic Relationship Dimension to Established Theories In relation to established theories, an emphasis on the therapeutic relationship as central is consistent with psychodynamic and phenomenological theories, although these theories may vary in the extent to which conceptualizing the relationship as a means or an end is consistent with the theoretical orientation. In psychoanalysis, for example, the transferential relationship is absolutely necessary, but it is the process of interpretation and related insight that actually creates change. Object relations theories focuses on creating a "good enough" attachment relationship to correct the developmental impasses from early child/parent attachment relationship.

Emotion-focused therapy (Elliott & Greenberg, 2007) and gestalt therapy (Perls, 1973), which are phenomenological, are also consistent with viewing the relationship as central, but also as the means, not the mechanism of change. A variety of techniques are used for experiential processing or accepting responsibility, which are the bases of change. In contrast, viewing the relationship as central and as the actual mechanism for change is more consistent with person-centered therapy (Rogers, 1951; Rogers et al., 1967), which is also phenomenological.

Seeing the relationship as neither central nor sufficient does not mean that you think that the relationship is completely inconsequential. Remember, we are starting with the acknowledgment that a good working alliance—the foundation relationship—is important to all psychotherapy. Most cognitive, cognitive-behavioral, and behavioral approaches are consistent with viewing the relationship as not central and as only a means in as much as a working alliance is necessary for the creation of trust, empathy, and positive regard as a basis for working towards the goal of client change. An exception to this is the post-modern cognitive constructivist theories, which are more likely to see the relationship as central and as a means to shifts in meaning making which are the actual mechanisms of change.

Feminist and multicultural therapists stress an egalitarian, collaborative relationship fostering social activism and institutional change. This relationship is seen as central, but also more as a means to the end rather than the explicit mechanism of change.

Real–Unreal Relationship

The real-unreal relationship dimension is related to whether the therapist emphasizes the transferential relationship or the here-and-now relationship between the therapist and the client or some mix of the two. In transference, a client re-enacts or displaces patterns, feelings, and attributions from prior relationships onto the relationship with the therapist. Some transference likely happens in most relationships of any sort as our impressions, feelings, and expectations about current relationships are affected by past relationships. But transference is conceptualized here as more than influence on current relationships, where the past is obscuring the present, actual relationship. The "unreal" relationship also includes attention to countertransference, which is the therapist's transference about the client. Countertransference is not the same as all feelings about the client. Therapists, just like clients, may have feelings and reactions to clients that are based in the real relationship.

In contrast, the "real" relationship is:

> that dimension of the total relationship that is essentially nontransferential, and is thus relatively independent of transference... The real relationship is seen as having two defining features: genuineness and realistic perceptions. Genuineness is defined as the ability and willingness to be what one truly is in the relationship—to be authentic, open, and honest. Realistic perceptions refer to those perceptions that are uncontaminated by transference *distortions* and other *defenses*. In other words, the therapy participants see each other in an accurate, realistic way (Gelso & Carter, 1994, p. 297, emphasis added).

Feelings and interactions based in the real relationship are related to the actual interaction and characteristics of the client and the therapist in the current moment and situation. Of course, the line between these two aspects of relationship is not as clear as the definitions might seem. A real interaction is interpreted by both parties in ways that are shaped by prior relationships as noted before. However, in terms of conceptualization and therapeutic intervention, the question becomes "Which aspect is emphasized?"

EXERCISE 4.15 ■ Describe your relationship with your intimate partner or your closest friend. Consider what you like about that person, the things that bother you about that person, how you act with that person, and what you like about yourself when you are with that person. Now, consider how your relationship with that person is similar to or different from your relationship with your primary caregiver. How is your partner/friend similar to your caregiver? Can you think of moments between you and your partner/friend that were less about what was between you and more about old patterns and relationships from your childhood? Was this easy or difficult for you to do? What was easy or difficult about it? The point here is not to suggest that your current relationship is "unreal," but to demonstrate how we might *conceptualize* the complex interactions of real and unreal aspects of relationships. Consider also whether there is such a thing as a true "real" relationship. Is this a function of development, that is older people might be more honest in their desires, and so on?

As in most of the other dimensions, there is no right or wrong answer. A therapist who emphasizes the transferential ("unreal") relationship will conceptualize the client and interactions with the client in terms of early relationships. For example, if the client reacts in a way that makes it clear that he or she has misinterpreted the therapist's intention, the therapist would wonder what early relationship was being mirrored. In contrast, a therapist who emphasizes the real relationship might see that misinterpretation as a breakdown in communication and wonder what the therapist had done to miscommunicate and also explore the client's current understanding.

Many therapists focus on both the "real" and "unreal" relationships, depending on the circumstances. Every relationship we have has its "real" and "unreal" aspects. In therapy, we can explore the "unreal" aspects and learn to differentiate between them and the "real," using this understanding of the difference to contribute to therapeutic change. In the discussion of Maria, above, the influence of the relationship related to both real aspects, such as both being women of color with felt obligations to honor their cultural communities, and to unreal aspects, such as the transference of Maria's expectations of self-sacrifice from early relationships to the current relationship with the therapist.

Milo, a 12-year-old Latino boy, was placed in foster care when his mother was arrested for armed robbery. His father was already serving a jail term, and he had been living with his mother and maternal grandmother when the arrest occurred. His school had sent Milo to a therapist at a neighborhood clinic two months before the arrest; the presenting issue was acting out in the classroom and anger management. Milo's therapist was a divorced young male who had some Latino heritage. The therapeutic relationship was intense; Milo yearned for his sessions and was very attached to his therapist who, in turn, was very fond of him. His teachers and grandmother noted some improved behavior. Milo began to indicate to his therapist—in his drawings and verbally—his desire to be adopted by his therapist, to become his son. The therapist, mourning his divorce and lack of children and family, expressed to his supervisor that, while he realized this could not happen, he wished at some levels that it could. The therapist required a great deal of supervision and support to deal with his and Milo's raw neediness. At the time of his mother's arrest, Milo began to cut himself; his neediness became understandably intense and he desperately wanted attention from his therapist. The therapist, who was only two months away from ending his internship, was able, with the help of his supervisor, to help Milo understand the genuineness of their attachment and also the differences between the "real" and "unreal" relationship. The goals of therapy became to stabilize Milo's living arrangements and to prepare him for termination with this therapist and transfer to another. This poignant example serves to help us understand how carefully we need to attend to all aspects of the therapeutic relationship by understanding our own needs and wishes as well as those of the client's.

In addition, there may be times when a client elicits feelings in the therapist that the therapist uses to point out to the client some transferential elements. Danielle, age 38, came to therapy because she was unhappy in her marriage. She externalized all of her feelings and blamed others for her not being "happy." She came to each therapy session with a litany of complaints about her husband, her friends, her co-workers: they did not smile enough at her; they did not return her calls fast enough; they did not ask her about herself enough. As with Nancy, the therapist found herself annoyed at Danielle's refusal to talk about herself and wanting others to accommodate to her wishes. When Danielle accused her therapist of not being supportive of her complaints, the therapist gently shared with her how she found it difficult to listen to the details of what others are doing wrong and how she wanted to help Danielle focus on what she had the power to change—herself rather than others. She wondered aloud if Danielle's friends also found it difficult to listen to the complaining. This was difficult for Danielle to hear but she did stop and reflect, and her tone and affect changed as she asked her therapist to say more about this. In further therapy sessions, Danielle began to consider whether or not her expectations of how others should treat her were reasonable.

A therapist needs to carefully consider if and when to use dynamics in the therapeutic relationship to call attention to interpersonal or transferential aspects of the client. It is difficult for anyone to hear that his/her actions are different than his/her intentions (a kind of confrontation); it is unlikely that Danielle actively wanted to annoy her therapist and, in fact, she was likely trying to elicit sympathy. It is also difficult because these kinds of interpretations can create feelings of conflict in the therapeutic relationship. In deciding to share her impressions with Danielle, the therapist carefully evaluated the strength of the working alliance. She also carefully considered whether Danielle would be able to tolerate the possible feelings of conflict, or dissonance about her own self-image.

Although some of the dynamics with Danielle were similar to those that the therapist initially felt with Nancy, the relationship and the basic conceptualization of the clients were strikingly different. The therapist felt that Nancy was much more fragile and that this kind of interpretation could be damaging. She also felt that Nancy's blaming of others was more related to self-protection, and thus this kind of interpretation would not actually be addressing the problematic dynamic. In addition, Danielle, as an older woman, had more experience and confidence in herself, which also related to this behavior being relatively more entrenched (and therefore, problematic).

In another case, Ben, a 15-year-old adopted African American boy, was referred to therapy by the school counselor. Ben's adopted parents were White European American and Ben attended a middle school with a diverse racial student body. Ben was teased by other students and taunted for having White parents. He developed a strong attachment to his male therapist and agreed to family sessions where adoption and racial identity issues were openly discussed. About two weeks before the end of the school year when the therapy was to terminate, Ben asked his therapist, "May I ask you a personal question?" The therapist, suspecting what the question would be, responded "Yes, if you tell me why you want to know and then what my answer means to you." Ben agreed, saying that he wanted to know in order to understand better about himself; he then asked the therapist, "Are you gay?" The therapist replied "yes" and Ben then said "You mean someone who is gay can become a doctor? Like me." Therefore, the last two sessions focused not just on termination, but also on issues of sexual orientation. In this instance, the therapist's self-disclosure when asked was validating, authentic, and turned out to be significantly therapeutic. This might not be so in all cases.

Lindsey was another client who asked the therapist about sexual orientation. She was openly lesbian and in an abusive relationship. In this instance, the therapist did not agree beforehand to answer the question, but inquired about what the question and possible answers might mean to Lindsey. Lindsey responded by saying that if the therapist were a lesbian, then she and Lindsey could have a sexual relationship. For Ben, this question (and its answer) was firmly situated in the real relationship, with the therapist being a role model. For Lindsey, the question (and its possible answers) were related to the unreal relationship, to Lindsey's fantasies and desires about what she wanted the therapist to be, regardless of the actual real relationship and the ethical boundaries that this relationship entailed.

Therapists need to consider whether self-disclosure is going to be helpful to the client, how and why. (We will talk more about this in Chapter 5.) One major criterion is whether questions demanding self-disclosure relate to the real or unreal relationship and the therapist's conceptualization of this.

EXERCISE 4.16 ▪ In small groups, discuss what kinds of transference issues might emerge with Nancy. How might the "unreal" aspects of the relationship affect your relationship with her? Consider also what your countertransference issues might be with Nancy. How can you and your peers help each other with these issues?

Relation of the Real-Unreal Relationship Dimension to Established Theories In relation to established theories, a relative emphasis on real or unreal relationship aspects does not mean that one cannot or does not recognize the presence of both aspects. And most therapists will attend to real or unreal aspects of the relationship that are actively affecting what they understand to be the process of change. But some theories (and therapists) emphasize actively working with different relationship aspects as ways to create therapeutic change.

Psychodynamic theories place an emphasis on transference and countertransference aspects of the relationship. In contrast, phenomenological/humanistic theories emphasize the real relationship. Family systems theories behavioral and cognitive-behavioral therapists may consider the role of past and current relationships in shaping cognitive and relationship patterns and behavioral contingencies. However, although these theories may consider the role of past relationships in the development of the presenting problem, the relative emphasis in sessions is on the real relationship, if the relationship is explicitly addressed. In most cases for behavioral and cognitive behavioral therapies, the relationship between the therapist and the client (whether real or unreal) is not conceptualized as a significant direct mechanism for therapeutic change as noted above.

Feminist and multicultural theories consider past power relationships as influencing current relationships and problems and use them as spring points for the deconstruction and reconstruction of more egalitarian and assertive relationships. While seeing the influences of past relationships, the relative emphasis is on the real relationship, as these therapies emphasize validating clients' experience and empowerment. These therapies object to the ways that focusing on the "unreal" relationship places the knowledge and power in the therapist, because it is assumed that it is the therapist who is able to see and interpret the transference.

Process Emphasis

Process is the relational meaning of interpersonal interactions, that is, what we are communicating about the relationship through out interactions. Yalom (2005) contrasts process with content and highlights that process is concerned with the interaction of intent and effect in the space between two individuals:

> [T]he *content* of that discussion consists of the explicit words spoken, the substantive issues, the arguments advanced. The *process* is an altogether different matter. When we ask about process, we ask "What do these explicit words, the style of the participants, the nature of the discussion, *tell about the interpersonal relationship of the participants?*" Therapists who are process-oriented are concerned not primarily with the verbal content of a client's utterance, but with the "how" and the "why" of that utterance, especially insofar as the how and why illustrate aspects of the client's relationship to other people. Thus, therapists focus on the metacommunicational aspects of the message and wonder why, from *the relationship aspect*, an individual makes a statement at a certain time in a certain manner to a certain person….identifying the connection between the communication's actual impact and the communicator's intent is at the heart of the therapy process. (Yalom, 2005, p. 143, italics in original)

Process involves the therapist taking all aspects of the apparent and underlying messages (verbal and nonverbal, direct and indirect) and putting them together in an attempt to formulate a response indicating empathic, deep level understanding. It is as if the therapist's eye and ear pick up elements of the communication that the client is not aware of and change is influenced as the therapist's conceptualization of his or her understanding is communicated back to the client.

EXERCISE 4.17 ▪ Consider the ways that different process messages may be conveyed. For each of the examples below, think about how you would react to each scenario. How would you feel? What would be your understanding of the other person's feelings? What is the other person trying to communicate to you about his or her feelings about you, your actions, or the relationship?

a. One night, you and your partner have a disagreement about where to go on vacation because you are really a beach person and your partner likes to explore cities. You go to bed angry at each other, which rarely happens. In the morning when you wake up, you are still somewhat angry, but really believe you can find a place to go where you can both be happy. You brush your teeth and go out to the kitchen for breakfast. Unfortunately, you have a habit of forgetting to put the toothpaste cap back on the tube of toothpaste and your partner has a strong preference for keeping the toothpaste capped. Consider the following scenarios that might occur when your partner comes out for breakfast:

—Your partner says, "I know it's hard to remember, but I'd really appreciate it if you could try to cap the toothpaste."
—Your partner makes a joke about getting set in our ways and senile as we get older and says, laughing, "like the toothpaste cap!"
—Your partner says angrily, "You forgot to cap the toothpaste *again*! You *know* how much it bothers me!"

Imagine different tones and body language for each of these responses and consider what these nonverbal signals would add to the process communication.

b. You have been working for three months with a client who came to therapy distressed because her girlfriend had recently broken up her after several months. In the past, she had had several other intimate relationships where women broke off the relationship when she desired to continue. In the previous session, you had let this client know that you would be taking a two-week vacation at the end of the month (you have two more sessions before you leave). The client arrives 10 minutes late for your session.

Consider the following scenarios:

- The client begins the session stating that she has decided that she, too, deserves a vacation and will be missing the next session (the last before you leave). She describes how she felt inspired by you to think about how to take better care of herself. She is planning a vacation with some other single friends.
- The client begins the session by stating that she will be missing the next session and wishes you a good time on your vacation.
- The client spends the session talking about how angry she is at her ex-girlfriend and sharing different fantasies about "getting back at her" in various ways. At the end of the session, as you walk the client to the door, she tells you she will be missing the next session.

Again, imagine different tones and body language that might accompany each of these responses and consider what these nonverbal signals would add to the process communication. How would these scenarios be different if you had not just informed your client that you were leaving on a two-week vacation? How would they be different if the client had come to therapy because he or she had a plane phobia?

A process emphasis would consider not only the content of the communication (for example, "please cap the toothpaste," or "I'm going to miss next session") but also why the person chooses to communicate in the way he or she and at the time that he or she and how this is communicated. Process emphasis is a relational variable because it is not only about what the client (or other individual) consciously intends, but also about what the effect might be on the other person, and on the iterative loop of interactions.

For example, you might respond to the last scenario about the toothpaste cap by asking "Why are you so angry about it?" and your partner might say "I'm not angry. I just want you to cap the toothpaste!" But the effect of the process is to communicate anger. Perhaps, your partner is feeling that the toothpaste cap is another sign that you feel your needs are more important than his or hers (like the vacation fight), or that you do not care about him or her, or that you are, yourself, angry and left the cap off deliberately to annoy him or her!

Although we have tried to illustrate it here with simple single statements, accurate perception of process is highly contextualized and develops over time. Process communication is rarely captured in a single statement, but rather in considering a sequence of statements or interactions, within a particular context (for example, the fight the night before, your two-week vacation) for a particular person. If your partner was obsessively neat, or always irritable before coffee in the morning, or currently concerned with getting older, the meaning of the process in each scenario might be different.

Process illumination is bringing the process into the client's conscious awareness as a means to create change. Yalom (2005) describes it as helping the client recognize and understand the following sequence:

a. Here is what your behavior is like.

b. Here is how your behavior makes you and others feel.

c. Here is how your behavior influences the opinion others have of you.

d. Here is how your behavior influences the opinions you have of yourself.

The therapist's awareness of process and his or her decision to communicate that awareness to the client are not the same thing. Most experienced and effective therapists have an awareness of process, but may rarely use process illumination with the client as a means to change. Overall, process and process illumination focus on the extent to which the therapist emphasizes the relational meaning of an interaction, attends to the multiple possible relational subtexts of interactions, and makes choices about whether or not to make these subtexts explicit as a strategy for therapeutic change.

EXERCISE 4.18 ■ Think about an interaction with a client (or a peer if you have not yet started seeing clients) that made you feel confused or uncomfortable. What about the interaction made you feel that way? Was it what was actually said or was it the process? How did the process meaning get conveyed? How aware of this message were you? How did you react at the time? Considering now the issues of process, how might you react?

Relation of the Process Emphasis Dimension to Established Theories In relation to established theories of therapy, therapies that see the relationship as highly important are more likely to explicitly attend to process dynamics, although therapists from all orientations might attend to process to further the therapy goals. Therapists from different established orientations vary much more in the extent to which they explicitly use process illumination.

Psychodynamic therapists frequently use process illumination with an emphasis on the unreal (transferential) relationship. Behavioral and cognitive-behavioral therapists may use process illumination but conceptualize it more in relation to learning theory or cognitive distortion and less in relation to relational insight. For example, "here is what your behavior is like" may relate to becoming aware of behaviors and cognitions through observation or daily journaling; "here is how your behavior makes others feel and judge you" may relate to the effects of behaviors; "here is how you judge yourself" may relate to cognitive evaluations of self. But the mechanism for cognitive-behavioral therapies is not only the understanding or insight through interpretations of these process meanings, but more particularly the active change in current behaviors or cognitions.

Humanistic and phenomenological therapists (including existential therapists such as Yalom) pay significant attention to process, but may vary in their use of or approach to process illumination. Person-centered therapists, for example, may utilize their own process awareness in conceptualizing the client as part of a genuine relationship, focusing on understanding a client's behavior, the effect on the therapist as an authentic response, the consideration of feelings about the client within a framework of unconditional positive regard, and consideration of how a client's behavior may contribute to feelings of inauthenticity. Existential, gestalt, and other phenomenological therapy orientations may be more likely to use process illumination directly (as Yalom describes) or as a part of other techniques aiming to increase the client's awareness of his or her own experiences internally and interpersonally.

Systems therapy is very consistent with focusing on process understanding and utilizing process illumination as one way to approach changing relational interactions and systems. This approach uses role-plays, direct feedback, and re-enactments to provide the feedback loops that restructure and reorganize the system relationships.

Feminist and multicultural therapies are frequently very attuned to the ways that systemic inequities of power and privilege affect process. For example, the communication patterns of White European American men, discussed in Chapter 3, may be understood to have a process meaning of "my voice is more important than yours" because of the socialized power and privilege associated with White male status. Feminist and multicultural therapies may help clients see the ways in which their own and others' relational process is related to sociocultural systems and help clients use this awareness to become empowered for positive change.

INTERACTIONS OF RELATIONSHIPS AND LOCATION OF KNOWLEDGE

In Chapter 2, we briefly discussed your epistemological philosophy and whether you think that the therapist or the client has the primary knowledge necessary for positive change. Now that we have explored more of your beliefs about people, contexts, and relationships, we want to briefly return to these questions and consider the implications of your epistemological philosophy on your understanding of clients and change processes. As you may recall, we introduced the question of whether the knowledge necessary to create positive change is primarily in the client or the therapist. One could see this question as another dimension of therapist preference. However, we see it as a foundation that affects many dimensional preferences. We understand it that way because the location of knowledge for change reflects a broader understanding of how knowledge is created and validated. We believe this affects all of the other dimensions because it affects your view of whether it is possible to have a understanding of truth, reality, or health that can be applied to all people or whether these understandings look different for different individuals, groups, or societies.

If you more strongly believe that there is one right or normal way to think, feel, or behave, then you are more likely to evaluate health and pathology in relation to what is agreed upon as healthy, as established by norms, social standards, or experts. Your evaluation of different contexts is more likely to be evaluated in relation to this idea, rather than in relation to the functionality for a specific individual in particular interactive contexts. You are also more likely to believe that you have more knowledge than the client about what would be most functional or healthy for them because of your expert training. This is likely to influence you to be more directive, and possibly more active, structured, and confrontative. If, on the other hand, you believe that knowledge is created and validated within particular people in specific contexts, you are more likely to evaluate heath and pathology in relation to the specific circumstances of the individual. You are also more likely to believe that, although you have expert knowledge and understanding about people and change processes generally that you can offer to the client, the client has the necessary knowledge for change, because this is always unique to the individual.

PULLING IT ALL TOGETHER

In the last three chapters, we have encouraged you to explore your own beliefs about change, contexts, and relationships. This awareness will help you develop a personalized theoretical orientation (ideas about change) that will be the foundation of a case conceptualization for any specific client. In the next chapter, we will discuss some dilemmas that novice therapists frequently encounter. We will then move into the next section, which focuses on the process of developing a case conceptualization. But before we move on, let's step back and try and get an overview of your awareness of your beliefs and preferences.

EXERCISE 4.19 ■ Rank the following statements and then compare your answers with those of another student. The first eight statements are about your understanding and awareness. The next 15 statements are about your comfort with different approaches.

Rank these statements 1 (not at all true), 2 (not really true), 3 (somewhat true) to 4 (absolutely true).

1. I have a good understanding of how I have made changes in my own life when I have wanted to do so.
2. I have a good understanding of my own values and worldview.
3. I have a good understanding of my cultural values and background, and how my experiences with social systems of power and privilege have shaped who I am and what I believe.
4. I have a good understanding of when and why I think something is healthy or pathological.
5. I have a good awareness of my own interpersonal style, whether I prefer being active or not active, directive or not directive, process structured or not process structured, confronting or not.
6. I have a good awareness of my understanding about how the therapeutic relationship affects change for the client, that is, whether the relationship is central, whether it is a means or an end to change, and my views about the importance of attending to relational process.
7. I have a good awareness of the ways that social structural and role power influence me generally and influence my therapeutic encounters. I can integrate this analysis into my understanding of the client.
8. I have a good understanding of what a working alliance is and how it differs from a relationship where people like each other.

For the next group of statements, think about your experiences as a therapist (or imagine yourself as a therapist, if you have not yet begun your practice). Rank these statements from 0 (very uncomfortable) to 4 (very comfortable).
I am (very uncomfortable, uncomfortable, comfortable, very comfortable) with…

9. Identifying the client's "problems" and critically questioning and examining why I think these are problems.
10. Considering the client's present experiences and relationships and how these affect the client's functioning (positive and negative).
11. Considering the client's past experiences and relationships and seeking connections between past and present in order to understand how these affect the client's functioning (positive and negative).
12. Considering the client's experiences and aspects of worldview that are different than my own (for example, cultural experiences, family experiences).
13. Considering the client's experiences with discrimination or oppression, even if it is from people like me.
14. Considering how this particular client might benefit from me as a therapist being more or less active, directive, structured, or confronting.
15. Considering how the ways that the client thinks affects his or her functioning (positive and negative).

16. Considering how the ways that the client feels affects his or her functioning (positive and negative).
17. Considering how the ways that the client acts or behaves affects his or her functioning (positive and negative).
18. Considering how the client is in relationships and how he or she communicates relational messages, including the relationship between me (the therapist) and the client (process-oriented).
19. Considering the "unreal" relationship: the client's feelings of transference and my own feelings of countertransference.
20. Considering the ways that the client's family may affect the development, maintenance, or change of presenting problems and of strengths and resources.
21. Considering the ways that the client has made personal choices that affect the development, maintenance, or change of presenting problems and of strengths and resources.
22. Considering the ways that the client's community or social networks may affect the development, maintenance, or change of presenting problems and of strengths and resources.
23. Considering the ways that the client's social context and statuses (for example, race, social class, sexual orientation) may affect the development, maintenance, or change of presenting problems and of strengths and resources.

Exercise 4.19 focuses on your awareness of your own views and your comfort in considering different aspects of the client's experiences and of the therapy. If you have rated any of the items very low, we would encourage you to think more about that area and try to develop greater clarity.

Your comfort and expertise in thinking about and interacting with clients will change over time. Novice therapists are frequently more comfortable with understandings and interactions that are closer to everyday social interactions or that require less expert knowledge, awareness, or skill. For example, they may be more comfortable with considering the impact of and exploring present experiences and real relationships than with considering and exploring the connections of past to present or the transference relationship. The latter requires more interpretation of experience and is frequently farther from the client's own awareness. This is one reason why it is important to understand your theoretical orientation more generally. Your theoretical orientation may be more abstract and may reflect ideas and approaches that you are not yet fully comfortable with but would like to work towards developing.

EXERCISE 4.20 ■ Consider the multiple dimensions we have explored. Where do you position yourself in relation to these dimensions?

	Contextual Dimensions
Location of Problem	Individual, Family, Socialstructural
Focus of Change	Cognitive-Affective-Behavioral
Past–Present Emphasis	Past vs. Present

	Relational Dimensions
Directive vs. Non-directive	Directing or specifying content vs. letting client choose content
Structured vs. Unstructured	Structuring how content is explored (e.g., through activities, exercises, etc.) through letting client decide (usually open discussion)
Activity Level	How active the therapist is, for example, how much the therapist talks or shares with the client
Confrontativeness	Extent to which the therapist calls the client's attention to contradictions, discrepancies, or things of which the client may be unaware
Significance of Relationship	This dimension addresses the extent to which the therapy relationship is seen as important for the facilitation of change. It includes the centrality of the relationship, which addresses whether the relationship—or the particular nature of the relationship beyond a working alliance—is seen as necessary for facilitating change. If the relationship is seen as central, then this dimension also addresses whether the relationship is seen as directly facilitating or creating change, or as an important means to other mechanisms of change.
Real Unreal Relationship	Emphasis on real relationship vs. transferential relationship
Process Emphasis	Extent to which the therapist attends to the process dynamics (process awareness) and uses this awareness (directly or indirectly)

Now consider how comfortable you would be conceptualizing (thinking about) or interacting with (treating) a client differently than your preferred approach? How flexible are you in tailoring your theoretical orientation to a particular client? The answer to this question will relate to issues of fit between you and particular clients, which we explore further in the next chapter.

EXERCISE 4.21 ▪ What is your theoretical orientation? Try and describe your own beliefs in your own words, rather than simply identifying your agreement or disagreement with established theoretical orientations or with the dimensions above. How do people develop problems or develop in healthy ways, how do they change, and how and why does therapy work? Why do you believe as you do? How will your beliefs contribute to the creation of change? Remember that your theoretical orientation will likely become more sophisticated and complex as you develop more experience as a therapist. This is only your current understanding and you are not expected to know everything.

Becoming an expert therapist is a developmental process. In earlier years, your understanding of clients is likely to be closer to your understanding of yourself, or to the theories you are most exposed to in your training and supervision. As you develop, your understanding of how people develop and change will expand and deepen, as you learn from your clients, from your successes, and from the challenges you encounter as a therapist. You will also learn new techniques and ways of enacting your understandings. Thus, the range of responses available to you will broaden, and these interactions will inform your understandings of clients as well. We are suggesting that there is a parallel process in client development and change with clinician development and change. As we continuously work with clients to reach target goals of desired change, we need to attend to our own target goals as well.

EXERCISE 4.22 ▪ Write a paragraph explaining your ideas of what case conceptualization is and what it means to you. Write a second paragraph describing how you think your theoretical orientation will affect your case conceptualization. Exchange your paragraphs with a partner and then discuss your thoughts about your own and your partner's paragraphs.

SUMMARY

In this chapter, we have explored your understanding about the nature of the therapeutic relationship and its influence on change. We ended with an overview of your explorations of your own worldview and ideas about change, your thoughts and awareness about client contexts, development, and focal points for creating change, and your views about therapeutic relationships. In Chapter 5, we will turn to an exploration of commonly encountered dilemmas before moving on to considering how the understandings and awareness discussed in Chapters 1 to 4 can be integrated with information about particular clients to create a conceptualization of clients to guide treatment planning and intervention.

REFERENCES

Elliott, R. & Greenberg, L.S. (2007). The essence of process-experiential/emotion-focused therapy. *American Journal of Psychotherapy*, 61(3), 241.

Gelso, C.J. & Carter, J.A. (1985). The relationship in counseling and psychotherapy. *The Counseling Psychologist*, 13(2), 155.

Gelso, C.J. & Carter, J.A. (1994). Components of the psychotherapy relationship: Their interaction and unfolding during treatment. *Journal of Counseling Psychology*, 41(3), 296–306.

Gelso, C.J. & Hayes, J.A. (1998). *The psychotherapy relationship: Theory, research and practice*. New York, NY: Wiley & Sons.

Mallinckrodt, B. (2010). The psychotherapy relationship as attachment: Evidence and implications. *Journal of Social and Personal Relationships*, 27(2), 262–270.

Miller, W.R. & Rollnick, S. (2002). *Motivational interviewing: Preparing people for change* (2nd ed). New York, NY: Guilford.

Norcross, J.C. (2002). *Psychotherapy relationships that work*. New York, NY: Oxford University Press.

Okun, B.F., & Kantrowitz, R.E. (2007). *Effective helping: Interviewing and counseling techniques* (7th ed.). Pacific Grove, CA: Thomson Brooks/Cole.

Perls, F. (1973). *The Gestalt approach & eye witness to therapy*. Palo Alto, CA: Science and Behavior Books, Inc.

Prochaska, J.O., Norcross, J., & DiClementer, C. (1994). *Changing for good*. New York, NY: Avon.

Rogers, C.R. (1951). *Client-centered therapy: Its current practice, implications and theory*. London: Constable.

Rogers, C.R., Gendlin, E.T., Kiesler, D.J., & Traux, C. (1967). *The therapeutic relationship and its impact: A study of psychotherapy with schizophrenics*. Madison, WI: University of Wisconsin Press.

Yalom, I.D., & Lesczcz, M. (2005). *The theory and practice of group psychotherapy*. New York, NY: Basic Books.

5

Dilemmas in Effective Helping

Even if one carefully gathers information and tailors the conceptualization and interventions to a specific client, dilemmas can occur. Some of these dilemmas can baffle even the most experienced therapist. Novice therapists, however, may be taken unaware by various issues that relate to conceptualization and treatment planning. This chapter explores some of the dilemmas that frequently cause anxiety in novice therapists. There are not always clear-cut answers to the questions raised, but we want to at least consider the issues that might arise during therapy.

We are aware that this chapter cannot address all or even most of the various dilemmas that may emerge in therapy. Ethical guidelines provide the necessary foundation for beginning to consider how best to address these dilemmas, so we refer you first and foremost to the ethical and legal guidelines for your particular treatment context and discipline: American Psychological Association (APA); National Association of Social Workers (NASW); Association of Mental Health Counselors (AMHC); American Counseling Association (ACA); American Association for Marriage and Family Therapy (AAMFT); American Psychoanalytic Association (APSA). (See also state legal guidelines and any guidelines specific to your organization.) However, while you should always follow legal and ethical guidelines, ethical guidelines are not always enough, and your own ethical standards may be more stringent than official guidelines. For example, there is much controversy within the American Psychological Association organization and membership about the ethical stance of the APA on psychologists' participation in the interrogation of "detainees" post 9–11 at Guantanamo Bay (see references at the end of this chapter), which ultimately resulted in a member-generated resolution. This is an instance where a large number of psychologists felt that the organizational ethical standards were not enough, and worked to change the standards. Thus, it is important not only to be familiar with and follow ethical and legal guidelines as written, but also to explore and understand your own ethical standards and consider how you will address possible dilemmas and conflicts between your personal and professional codes of ethics as you make choices about conceptualizing and intervening with clients.

Ethics courses and books take up many of the issues related to ethical practice and exploring areas that are not as clearly defined where trainees may have questions (for example, Pope & Vasquez, 2010). Our purpose here is not to take up general ethical issues such as scope of practice, but instead to consider a few dilemmas that are particularly related to the issues of orientation, conceptualization, and treatment planning that we have been and will be discussing. In this chapter, we take up the following issues:

1. Fit of therapist and client and related issues of making decisions about referrals to adjunct care providers or a different therapist.
2. Boundaries of the therapy relationship, including issues of therapist disclosure, dual relationships (particularly in small communities), and managing issues of required breaks in confidentiality.
3. Issues related to social justice and cultural competence, including questions of imposing values and ideas about mental health and dilemmas related to internalized oppression.

FIT OF THERAPIST AND CLIENT

The issue of fit concerns the complexity of intersections of the therapist and the client. With training, experience, and supervision, you will not only learn to identify your areas of comfort and discomfort, but you will also be able to gain some understanding of *why* you experience differing levels of comfort and discomfort with different clients. This may relate to client characteristics or contexts, to the nature of the problems that the client presents, and/or to some of your own personal experiences or values. It is important to be aware of the issues and people with whom you are more or less comfortable and to explore the reasons why. If you are seeing a client with whom you are uncomfortable, this discomfort will likely affect your conceptualization as well as your interactions with the client; your own discomfort may contribute to understanding a client's behavior as more pathological, locating the problem in the client and being less sensitive to contextual influences, being more confrontative, or conceptualizing relationship issues as transferential when they may actually be reflecting aspects of the real relationship as the client picks up on your discomfort. Be aware, too, that your comfort zone will change as you become more skilled and experienced and as you change as a person, not only as a therapist. Throughout your professional development, it is likely that you will expand your comfort zone and be able to work successfully with a greater variety of people, problems, and contexts.

EXERCISE 5.1 ▪ Consider the following scenarios. Which would you feel more or less comfortable working with as a client? What aspects would you be more or less comfortable with?

a. John is a 16-year-old Latino who has been suspended from high school for bringing a weapon to class. He is a first generation immigrant, lives with his mother and younger siblings in a low-income housing project and is seen as a "troublemaker" in his community and at school. Although John is several grade levels behind in verbal and reading skills, he is currently in the 11th grade, having been passed through each year in spite of his limited academic skills. The school psychologist diagnosed John as having "conduct disorder with a possible antisocial personality." Her notes indicate that she believes he is involved in a gang. He expresses his rage at being mandated to undergo counseling and refuses to engage with you.

b. Tara is a 26-year-old White European American, referred to you after four weeks in alcohol rehab. Her husband is in the military and has been in Iraq for two years. Tara has a history of anorexia and has been drinking since early adolescence. A trained nurse, she was on family leave to care for her 3-year-old daughter and was remanded to rehab after a DUI with her daughter in the car.

c. Emily is a 28-year-old third generation Chinese American woman. She has been hospitalized several times for major depressive episodes and twice for attempting suicide. She is studying psychology at the local community college, but is currently fearful that she will have to drop out because she is feeling increasingly out of control. When she feels out of control, Emily cuts and scratches herself with safety pins, knives, or razor blades. She is clear that she is not, at these times, feeling like she wants to commit suicide or attempting suicide. Her prior records state that she is "manipulative," using the self-injury behavior to punish therapists whom she perceives as unhelpful.

d. Tom, a 61-year-old African American, lost his job as a computer programmer and has been out of work for eight months. His wife is a teacher and now the sole family earner. Their children are grown and self-reliant. Tom is experiencing age discrimination in his job search and has become morose and withdrawn, staying home all day in his pajamas watching TV, gaining weight, and sleeping restlessly. His physician diagnosed him with depression, prescribed an SSRI and referred him for counseling.

e. Frank is a 41-year-old White European American man who is mandated to be in therapy after being arrested for child sexual abuse. In the course of the first few sessions, it becomes clear that he has abused multiple children ages 2 to 10. He is an active member of an association that promotes the "positive" aspects of love and sex between adults and children, and he speaks passionately about the need for greater acceptance and freedom for children's sexuality.

f. Ben is a 25-year-old second generation Haitian American veteran who served in Iraq. He was hospitalized due to injuries sustained from a thrown grenade, which killed two others in his squad. Although Ben has a permanent limp and scars to his face and body from that incident, his response to the intake worker who asked about that experience was "that was the least of it." He has intense nightmares and flashbacks and has been unable to find a job.

What are your thoughts and possible biases about the clients just described? See if you can get in touch with your areas of comfort and discomfort.

Client characteristics contributing to your comfort or discomfort may include: demographic characteristics, personality, presenting problems, life style, motivation and the willingness of clients to engage in therapy. Clients' cultural characteristics and contexts may also impact your level of comfort. The clients briefly described in Exercise 5.1 reflect a range of personal characteristics, presenting problems, and contexts. While you may not feel strongly uncomfortable imagining working with any of them, the relative comfort or discomfort raised within you by imagining working with these clients can contribute to your understanding of areas for growth and consideration. In general, when considering fit between yourself and a given client, questions you can ask yourself may include: How does my understanding of change and relationships interact with others' understanding of these things (for example, the understandings of your clients, your peers, your supervisors)? How does my understanding of change and relationships interact with the different issues clients may bring? What and how can I learn from the literature and other professionals to enrich my understanding? Understanding your own values, contexts, and understanding and style of relating (as discussed in earlier chapters) can help you answer these questions and consider how your comfort or discomfort may affect your conceptualization and interventions with a given client.

Novice therapists are frequently concerned that they will be perceived as or actually be insufficiently knowledgable or skillful. They may feel less comfortable with older clients or more educated clients because of this fear. Therapists who have never been married or who do not have children may feel concern that they will not have credibility with married clients/couples or with parents. There is nothing inherently wrong with feeling more or less comfortable with particular clients, but your feelings of comfort should not necessarily dictate your actions or choices. In some cases, it might be best for an older client to see an older or more experienced therapist, depending upon such concerns as the client's presenting problems and relational issues. For example, you will need to consider what experience you have with couples and families and how that affects the goals and contexts of your treatment with a particular client in order to evaluate whether or not this will be an impediment.

However, it has generally been our experience that this kind of fear is more frequently the trainee's fear, rather than a concern of the client. Working through that kind of fear can lead to growth for the therapist and very successful therapy for the client. Remember that the foundation of good therapy is empathy and perspective taking: you will never have had all the experiences that your clients have, even if you are 102! No one therapist can have personally experienced all of the issues a client may present; it is our ability to empathize and understand others' experiences that determines the level of our helpfulness.

While you may be concerned about seeing clients with experiences with which you are not familiar, your comfort with clients may also relate to your own particular experiences in relation to a client's presenting issues. When you conceptualize a case, you integrate external and internal information from both the client and yourself. You bring your own experiences to your understanding of the clients' experiences. Consider the following case:

> Cara is seeing a couple in therapy who are having marital difficulties and have very different goals for the outcome. The couple has been married for 32 years. Seven years ago, the wife suffered a stroke. About a year before that, their youngest child left home and the husband took a high profile, demanding position as a CEO for a large company. Currently, the husband wants a divorce, explaining that his wife "has a different personality since her stroke seven years ago" and he no longer loves her. The wife is devastated—she states that she feels he is abandoning her and that his abandonment began with her stroke.

EXERCISE 5.2 ▪ Consider possible ways that Cara might understand what is going on here. How might different experiences from Cara's own life affect her conceptualization? Consider how Cara might understand what is going on with this couple if her own experiences had included any of the following:

- Cara's parents divorced quite recently and her mother was devastated and lonely.
- Cara's own mother had a heart attack when Cara was a senior in college. Cara deferred her entrance into graduate school for two years in order to care for her mother.
- Cara's mother suffered from episodic major depressive disorder. When she was not depressed, she was attentive and caring of Cara. When she was depressed, she seemed like a different person, neglecting Cara and her siblings.
- Cara has a close relationship with both her father and mother. Her father owns and runs a small business and her mother was a homemaker throughout Cara's life. Both her parents have encouraged her ambition to become a psychologist,

and she feels that she has been shaped strongly by both the model of her father's business ambition and her mother's devotion to her family.

EXERCISE 5.3 ■ If you were seeing this couple, what experiences in your life might affect your conceptualization? Compare your answers with others in a small group. This true example has many loaded issues likely to elicit strong feelings and reactions involving your values and attitudes towards marriage, illness, divorce, and individual and relational changes over the lifespan. Would you feel differently if the couple had been married 20 years? 10 years? See if you can get in touch with some of your inherent assumptions about people and relationships.

EXERCISE 5.4 ■ Review the case of Nancy from Chapter 1. Consider how your experiences are similar to or different from Nancy's, and what experiences in your life might affect the way you understand and relate to Nancy.

Being aware of your own experiences in relation to your clients' presenting issues can help you separate your own struggles (and possibly unresolved issues) from your conceptualization of your client. Most of the time, an active awareness of how your own experiences relate and may affect conceptualization and intervention is enough to ensure a focus on the client. But sometimes this awareness helps you identify when your own issues may make it inadvisable to see a particular client at a particular time. Making the decision that the fit with a client is not good enough to continue does not mean that you are not a good therapist. If, for example, Cara's parents were in the current process of considering divorce, were intensely fighting on a regular basis and involving Cara in these fights, and their issues—like the clients'—were related to changes due to illness, then it would be understandable that Cara might feel too close to these issues to be able to separate her own current and unresolved experience from the clients' at this time. In this case, a referral would be the best option and Cara would be demonstrating the awareness and skill of a good therapist in recognizing this conflict.

Ethical guidelines require that you have the necessary knowledge and skills to work with the client. Such knowledge and skills may relate not only to the client's contexts, but also to the presenting problems. In relation to the presenting problem, many clinicians find that they are more comfortable and competent understanding and working with certain categories of problems than others, and many types of problems require specialized knowledge and skills. Most of us start out more as generalists and then become more specialized with time and experience. While "fit" issues about presenting problems are frequently more on a continuum of comfort and discomfort, there are some times when a particular type of skill or knowledge is indicated as needed for the best treatment for a given client. In these cases, it is important to recognize the limitations of your expertise.

Research suggests that some presenting problems are best treated with particular kinds of therapeutic interventions. An example is specific phobias, such as fear of flying, heights, or spiders, which respond well to cognitive–behavioral interventions involving anxiety reduction, desensitization, and exposure (Barlow, Allen, & Basden,

2007; Okun, 1990). It is, of course, important to consider not only the research and the diagnosis, but also the client's context, understanding of his or her problems, and readiness for change. It is these factors and not the knowledge of the therapist that should influence whether these strategies are used. In most cases, a therapist who is not skilled in these interventions will not be the best fit for a client presenting with a specific phobia as the primary area for intervention. In such a case, a referral is in order, either for adjunct treatment specifically for the phobia (if there are other major areas for which you *are* a good fit as a therapist) or as primary treatment. You may suggest to the client that you think a specialist would be more helpful for this particular issue and suggest a referral. You will need to explain your reasons for this referral and review with the client what you have achieved working together and how each of you feels about this referral.

Skills and knowledge relate not only to the presenting problem, but also to the client's multiple contexts. We need to be able to tailor our conceptualization and interventions to different clients with different contexts. One example of this is your knowledge about different cultural contexts. For example, if a Muslim Pakistani college-age woman comes in because she is unhappy with the candidates for an arranged marriage that her family has provided, you need to conceptualize her case and plan interventions from a culturally sensitive perspective and not attempt to impose individualistic cultural views on her. This does not mean simply condoning the arranged marriage, but it does mean recognizing that the unhappiness she is experiencing may be related to a cultural intergenerational conflict and her desire to be a good daughter conflicting with her lack of positive feelings about the current candidates. Or she may be unhappy with the candidates and not the idea of arranged marriage. If this client was from the dominant European American culture and presented with issues related to feeling her family was pressuring her to marry, your case conceptualization might be different (depending on her specific ethnicity, her family context, immigration status) in that you might choose to help her to individuate from her family and take a stronger role in seeking what she would consider to be an appropriate mate, attitudes and behaviors which are more culturally congruent with European American culture.

While no therapist knows everything about every cultural and social context, it is important to evaluate whether the depth and breadth of your knowledge (and the knowledge you can gather) is sufficient to treat this particular client. In the case described above, it may be enough to know general information about collectivistic and individualistic cultures, and different family norms and systems, and to seek out additional information about the particular cultural contexts relevant to the client's experiences, in this case, Muslim and Pakistani cultures and practices and arranged marriages. On the other hand, consider the depth and breadth of knowledge you might need with a transgendered client who is a lawyer specializing in gender and transgender-related rights suits, an activist in transgendered politics, and who is presenting with the need to explore ambivalence about seeking sexual reassignment surgery. This individual has expert knowledge and strong personal and political feelings about an area that most therapists know little about and that the field of psychology has (and, many feel, continues to) pathologized and oppressed. S/he might benefit from a referral to a therapist with more extensive knowledge about the personal, cultural, and social justice issues related to transgendered peoples, if you do not have this knowledge.

Another area of fit that can create dilemmas in psychotherapy is that of ideological differences. What we mean by ideological differences are differences in values or beliefs about what is moral, right, or healthy. Ideological differences are frequently present between clients and therapists; it is not necessary that you and your client agree on everything. But ideological differences can create major dilemmas when the presenting problem is related or becomes related to the area of ideology or when the difference leads to adverse feelings between the therapist and client. This is more frequently a problem when the therapist has adverse feelings or beliefs about the client or the client's beliefs. This is so because a) the therapist is more likely to know about the client's beliefs than the client is to know about the therapist's beliefs; b) the therapist has the power in the relationship; and c) the aim of the therapy is to help the client. Ideological differences are frequently related to issues that are debated in politics, media, and other social contexts, and that are emotionally "loaded."

Thus, an important question to ask yourself is: "What values or beliefs do I have that are 'absolute,' where I may have difficulty working with a client who differs from me in these particular values?"

EXERCISE 5.5 ■ Consider the following clients. Which would you feel more or less comfortable with as a client and why? What kinds of dilemmas might emerge?

- Mary is a 20-year-old White European American woman who strongly believes that abortion is murder and a mortal sin against God. She has been active in rallies, pickets, and blockades of clinics and, while condemning the violence, speaks admirably of those "warriors" who have destroyed clinics or threatened doctors who perform abortions.

- Lou is a 20-year-old White European American man who strongly believes that conception is not the beginning of human life and therefore supports abortion as a woman's choice. He has been active in rallies, pickets, and confrontations with those who have blockaded clinics and feels that those who are against abortion are irrational, hypocritical religious zealots who care nothing about women and children.

- Peter is a 40-year-old African American gay man. He came out at age 15 and describes his teen and early twenties as "a bit wild" with multiple, sometimes simultaneous sexual relationships. He met his current partner at age 30 and they married a couple of years later. They have two children, ages 2 and 4. Although he says he loves his husband, he describes feeling bored with the relationship, has

recently felt "down," and states that his relationship with his husband is not as satisfying as it once was.
- Simon is a 40-year-old African American heterosexual man. He had his first sexual relationship at age 15 and describes his teen and early twenties as "a bit wild" with multiple, sometimes simultaneous sexual relationships. He met his current partner at age 30 and they married a couple of years later. They have two children, ages 2 and 4. Although he says he loves his wife, he describes feeling bored with the relationship, has recently felt "down," and states that his relationship with her is not as satisfying as it once was.

In the above cases, we have not specified why the client is coming to therapy. How would your feelings about working with these possible clients vary depending on the reasons why they were seeking therapy? Imagine that Mary and Lou are both seeking therapy because they are experiencing intense anxiety about exams at school. Would your ideological similarities or differences affect the therapy or the relationship? Now, imagine that Mary is seeking therapy because she has been raped and is pregnant, or Lou is seeking therapy because he recently found out that his partner was pregnant and is refusing to have an abortion. Imagine that Peter and Simon are seeking therapy because they feel depressed and would like to address some feelings and beliefs that they feel get in the way of being intimate with their current partners. Now, imagine that over time, it emerges that both Peter and Simon attribute their depression to their sexual orientation. Their goal shifts from bettering intimacy to changing their sexual orientation and ending their current relationships: Peter is seeking conversion therapy (to make him heterosexual) and Simon is seeking therapy because he is feeling that he has denied his attraction to men for too long and is questioning his sexual orientation, his relationship, and his future.

Some of these issues relate directly to ethical mandates. For example, regardless of whether you believe that homosexuality is right or wrong, good or evil, natural or unnatural, it is unethical to practice conversion therapy, even if the client desires it (Anton, 2010). Sometimes there is no ethical mandate, but simply an ideological difference. Typically, in supervision we encourage trainees to continue with a troubling case and to learn about and work through their negative feelings *as long as it does not harm the client*. Usually it is not difficult to find something to connect with a person so that the work can proceed, but there may be times where a transfer or referral is in order to avoid harm to the client. These times are most likely related to "absolute" values that the therapist holds. Although trainees sometimes say that they can hold strong absolute values (for example, homosexuality is wrong) without imposing judgment on clients, we think this is an extremely difficult, if not impossible, task.

Furthermore, it can be very risky to accept a client because they are presenting with an "unrelated" problem. For example, Simon may seem to be presenting simply with a desire to strengthen his marriage and improve his relational skills. But it may emerge, during the therapy, that his feelings of dissatisfaction with his marriage are related to his deep and long abiding attraction to men. A therapist who is morally opposed to homosexuality may accept Simon as a client thinking that there are no troubling ideological differences. But what will happen when that difference emerges? How might the therapist's ideological stance affect his or her conceptualization of Simon and the subsequent interventions? How might Simon be affected by becoming aware of the therapist's stance or by a sudden referral for a reason that cannot be stated without risking harm to the client?

A final area of fit we want to address is similarities or differences in social statuses such as race, ethnicity, gender, sexual orientation, and social class. Some research suggests that matching clients and therapists on race, ethnicity, or gender may be beneficial (Karlsson, 2005), although overall the findings are mixed. Furthermore, even when benefits are found, it is not always clear whether this is because the therapists and clients share experiences that make understanding and connection easier, whether the benefit comes from the client's expectations or assumptions that there will be shared experiences and greater understanding, or from some other influences (such as language proficiency). Sometimes a client will request a match—for example, one of us saw a multiracial adolescent whose mother had requested, if possible, a multiracial therapist. In these cases, it is important to understand the reasons for the request, and the assumptions that might be made about the therapist because of a particular status. In this case, the mother felt that part of the young man's anxiety was related to feeling caught between the cultures and races of his two parents, and the mother assumed that a multiracial therapist would have a greater understanding of this than a monoracial therapist. In this case, the mother was likely right, given the therapist's own history of reflecting on being multiracial.

But sometimes these assumptions can be wrong or complicated by other factors. For example, a first generation Chinese American immigrant client may request a Chinese American therapist, assuming that they share cultural values and norms. But if the therapist is a fourth generation Chinese American, he or she may be very different than what the client expected. Ferguson and King (1997) describe exceptionally well some of the issues that arose in therapies as African American women therapists, with African American women clients. While matched in gender, race, and ethnicity, the clients and the therapists varied in social class. While the gender, race, and ethnicity matching suggested there would be certain similarities, the class differences were most salient to the clients.

In most settings, matching on variables other than gender is very challenging. What seems most important is not necessarily the exact match, but the necessary sensitivity, understanding, and ability to work with differences (Chang & Berk, 2009; also see review in Karlsson, 2005). Even when the presenting issues are directly affected by unmatched statuses, it is often the ability of the therapist that matters most. We know a trainee who is Asian

American who worked with White European American veterans who had fought in the Vietnam War. When some of his clients remarked on the fact that he "looked like the enemy," the therapist's ability to accept that statement and explore that experience for these clients was frequently an asset, rather than an impediment, to the therapy.

While differences can frequently be bridged, sometimes it is best to make a referral to another therapist. In another example, a Kenyan husband and American wife struggled over how to raise their biracial child and the husband refused to continue therapy with a White American female therapist. When the therapist felt that an impasse had been reached, she referred the couple to a male colleague from Uganda and the therapy proceeded more smoothly. One of the issues had been corporal punishment and the male African therapist was successful in helping the father to modify his expectations and parenting strategies. In this situation, matching both gender and culture was important.

Not every clinician can be effective with all clients or with all presenting problems. There may be times when a client would fare better with a therapist of a different race, gender, religion, or belief system, or with a therapist who has a different set of skills or specialization. It is the way we suggest a referral that determines its effectiveness. For example, you may decide that a young woman who has experienced abuse from the males in her life would benefit from a therapeutic experience with one of your male colleagues whom you believe to be empathic and supportive so that she will have at least one positive relationship experience with a male. You may suggest this directly to the client, sending a message that not all males are abusive and that she could benefit from a supportive, positive relationship with an appropriate male.

It is important to be careful about the way in which you communicate your reasons for referrals, particularly if the client is not actively requesting such a referral. Clients may fear that their problems are too overwhelming or so severe that only a "specialist" can help; they may experience feelings of rejection related to the referral. The way in which you communicate your reasons for the referral can, on the other hand, help clients feel that the referral is an indicator of your care for them and your belief that they can change. No one is an effective fit for all clients, which is why it is ethical and responsible to make a referral when you arrive at and have adequately understood this conclusion.

And there may be times when you just cannot like the client. Hopefully, you will be able to find at least one thing you like about the client that will help you to try to form a positive relationship with him or her in another session or so. If, after a couple of sessions, you are unable to develop an empathic positive regard or if the client informs you that he or she is not comfortable working with you, you might suggest talking about whether or not this is a "good" fit. After sufficient discussion, if you or the client decides not to continue, it is ethically imperative that you assist the client in finding a more suitable therapist.

BOUNDARY DILEMMAS

Therapy is a particular kind of relationship, with particular roles and responsibilities for the therapist. Ethical requirements are relatively clear about the unacceptability of some types of dual relationships, particularly intimate or sexual relationships. Relationships where the therapist may have power over the client (for example, a therapist being both a professor and a therapist) also need to be carefully evaluated for ethical implications. They are usually discouraged because the evaluative nature of the teacher–student relationship, for example, may detrimentally impact the therapeutic relationship. However, there are many "gray areas" related to boundaries in the area of dual relationships. For example, do you attend a client's wedding? Accept a gift or give a gift? Attend a client's family member's funeral? What happens when you attend a social event and find that a client is one of the guests?

There are no clear-cut answers to these dilemmas. Each situation is different and you need to consider your feelings as well as what it would mean to the client, and to understand these issues in relation to your overall conceptualization of a client and the impact of different actions on the change process. If a client asks you to attend his wedding, you might discuss openly with him all of the ramifications of what it means to him, what it means to the therapeutic relationship, to his family, and to other guests. This discussion, when integrated with your conceptualization, will help you understand the meaning to the client, and consider the decision that best facilitates the work and maintains both a positive relationship and the boundaries that you feel are most helpful. For example, you may decide to attend the ceremony but not the reception/party. When you run into a client at a social event, or the movies, or in a store, you can take your cues from the client about acknowledgement and then discuss whatever happens in subsequent sessions. Confidentiality demands that you not acknowledge clients before they acknowledge you, but it is important for the client to understand that this influences your actions.

EXERCISE 5.6 ▪ Consider the following scenarios. What issues in relation to boundaries and effect on the therapeutic relationship might arise? What choices should the therapist make in these situations? Which choices would you consider unproblematic and which would you be concerned about? Why?

- Cyndy is a Native American psychologist who is a professor at a state university in the Midwest and who also sees clients privately. She is active in her tribal community both socially at pow-wows and in relation to serving in various leadership capacities. She is the only Native American licensed psychologist seeing clients within an 80-mile area. She receives a referral from a friend and tribal elder within her community for Linda, a young Native American woman who is struggling with issues of depression and possible substance abuse. Linda is currently attending a community college majoring in psychology with the hope of transferring to the state university at which Cyndy works. However, she is currently considering dropping out of college, as she feels alienated and hopeless about her ability to fit in and achieve her goals. Linda has twice been seen by White European American therapists and dropped out of therapy within the first three sessions. Cyndy's friend feels that Linda strongly needs a Native American therapist who is connected to the community and is familiar with indigenous healing practices. Prior to referring Linda to Cyndy, she introduced the two women at a tribal social and Linda commented afterwards how comfortable she felt with Cyndy, contributing to the friend's decision to make the referral. Cyndy is trying to decide whether she should accept Linda as a client.

- Helen is a White European American Protestant therapist who has been treating Aaron, a 12-year-old White Jewish boy, for two years for anxiety and family issues related to his parents' divorce when he was 9. Aaron's parents initially hoped to find a Jewish therapist for Aaron, as their religion and culture are very important and they had concerns that Aaron might be rejecting his heritage due to bullying and name-calling at school. However, there was no Jewish therapist available at the local agency (where Helen worked) and Aaron's parents agreed to try working with Helen. All have been pleased with the arrangement and with the progress that Aaron has made. Recently, Helen received an invitation to Aaron's bar mitzvah and she is deciding whether she should go. Should she go? Should she give him a gift?

- Carol, age 54, is referred to a therapist by her friend, Judy, who is also a client of this therapist. The therapist has actually heard about Carol from Judy, both about Carol's life and relationships, as well as about her relationship with Judy. Judy mentioned to the therapist that she referred Carol. The therapist is considering whether she should talk with Judy about her reasons for referring Carol and possible concerns Judy may have about seeing the same therapist as her friend, whether she should accept Carol as a client (or even a consultation for referral to another therapist), and whether she should talk with Carol about concerns she may have about seeing the same therapist as her friend.

- Bob is a 35-year-old Asian American gay male who works at a college counseling center. While he is generally "out" to his friends, family, and colleagues at work and attends LGBTQ events, he does not explicitly come out to his clients and makes conscious choices to not reflect his sexual orientation in his dress, his public information, or office decorations. He has been treating Gary for several sessions. Gary is a 20-year-old multiracial White-Asian American undergraduate who came to therapy because he was questioning his sexual orientation. He had met a fellow Latino undergraduate man who identifies as bisexual and to whom Gary was strongly attracted. Bob is involved with a local gay organization and he is currently the leader in organizing a social event for gay men of color. In a session, Gary mentions that his prospective boyfriend has told him about this event and wonders whether he should go. Bob needs to figure out what he should say, both about Gary's decision and his own connection to the event and organization.

Your response to these boundary dilemmas will relate not only to the particular clients, and to the organization policies, but also to your own theoretical orientation and dimensional preferences. If you have a very strong psychodynamic or psychoanalytically influenced orientation, then you may be more cautious about dual roles; the non-therapy related role would intrude a "real" relationship on the transferential relationship which is the dimensional preference within this orientation. In contrast, if you were treating Judy for chronic pain from a car accident with biofeedback and a behavioral orientation, and Carol also had chronic pain from an injury, you may be more willing to see them both. Depending on your preferences on relational dimensions, however, you may or may not emphasize needing to talk with both of them about this. Through these examples, we encourage you to explore the limits of your boundaries as well as the influences on these boundaries and the relation of them to your theoretical orientation and dimensional preferences.

Disclosure

The example of Bob, above, brings up a related boundaries issue of therapist disclosure. How much should therapists share with their clients about their own experiences, statuses, or contexts? What should guide these decisions? To what extent should therapists actively be aware of what they are communicating about themselves? Are there experiences, statuses, or contexts that therapists should directly and actively disclose from the very beginning, either verbally to the client or through written descriptions of themselves or their practice? What should a therapist do if the client asks a direct, personal question?

EXERCISE 5.7 ▪ Consider the following experiences, statuses, or contexts. Which would you feel comfortable sharing with a client spontaneously (in which you, as the therapist, initiate the disclosure)? How would you address a direct question from a client about this issue?

- What you had for dinner last night?
- What kinds of movies or music you like?
- How long you have been practicing or how many clients you have seen?
- What kind of training for therapy you have had?
- Whether you have children, intend to have children, or want children?
- Your parenting or disciplinary practices?
- What race you are?
- What ethnicity you are?
- Whether you are an immigrant or where you were born?
- What is your sexual orientation?
- Whether you have had sexual partners of the same or different sex?

- Whether you have experience with particular sexual practices?
- Whether you are married or have a partner?
- What is your income?
- What social class you grew up in?
- What religion are you?
- Whether you have, yourself, been in therapy?
- What is your attitude towards LGBTQ people, People of Color, White people, religious people, atheist people, poor people? How much knowledge do you have about each of these?

Different orientations and different therapists have varying ideas about whether or how much therapists should disclose about themselves, their own contexts and statuses, and their experiences. These stances relate to a therapist's orientation and understanding of how therapeutic change occurs. For example, as noted above, psychoanalysis emphasizes the unreal transferential relationship as both a means and an end to change. Because of this, therapist disclosure is strongly discouraged, as the client is less able to project his or her internalized relationships if the therapist's "real" experiences shape the client's view and limit the client's projections. A contrasting example is feminist therapy, which emphasizes an egalitarian and real relationship and is much more accepting of therapist disclosure in general as a means towards establishing this kind of relationship. Your dimensional preferences will also affect your decisions about boundaries and disclosure. If you were Bob and had a strong emphasis on the sociostructural influence on problem development and resolution, you would be more likely to share your involvement in the LGBTQ organization and may, in fact, make different choices about disclosing your sexual orientation to clients more generally.

Therapists who adhere to an integrationist orientation (and even those who adhere to an orientation with explicit ideas about disclosure) frequently make decisions about disclosure based primarily on the client, the conceptualization, and the relationship. Even when this is the case, however, there are some issues that a given therapist would never or rarely feel comfortable disclosing. Which of the issues in Exercise 5.4 would fall in this category for you, at least at this time? Why are these issues "off limits"?

The most important issue in considering whether or not to disclose personal information is the effect it will have on the client and the therapy. Anticipating this effect is related to knowing the client well and conceptualizing them holistically. Sometimes clients are simply curious about their therapists. But more usually, particularly if the request for disclosure comes up mid-therapy, there is a reason for this curiosity that may contribute important information to your conceptualization of the client. Sometimes clients seek out information about their therapists because they are wondering whether the therapist can understand their experiences. If this is the case, the therapist needs to ask him or herself whether this is due to a general concern or fear about being validated and understood, whether this is due to a specific issue or concern (for example, having had negative experiences with other people or with other therapists so that the client wants reassurance, or needs particular expertise), or whether this is due to some other reason. Understanding the client's experience and perspective is the core of good conceptualization and good therapy. Even when a question seems very basic, a therapist should not assume that he or she understands why the client is asking the question.

Perhaps your first question to the client is something like "Can you tell me why you want to know this?" or "What will my answer mean to you?" There are times when the client's inquiry may be a matter of social convention, such as "Did you have a nice holiday?" or in response to the clinician's announcement of an upcoming vacation, "Where are you going?" There are no rights or wrongs about answering; you can say "I am going away." or name a specific destination. It all depends on the contexts and circumstances of the relationship. The important point is to communicate respect for the client's concerns and interests.

One of us, while a graduate student trainee, had a client who asked how many clients she had seen and whether he (the client) was her first client. The client knew he was being seen in a training clinic and that the therapist was a trainee. On the one hand, a client does have a right to know about the training background and expertise of the therapist. The therapist could have assumed that the question arose out of concerns about the therapist's qualifications and ability to help the client. However, rather than assume that this was the case, the therapist inquired about why the client was asking this particular question. The client responded that he hoped that he was the therapist's first client, as he would then be particularly special and the therapist would be likely to remember him forever. As the therapy progressed, the theme of being "special" and the client's fears that he did not matter to anyone (including the therapist) emerged as central. The reason for this initial question, then, became an important piece of information that contributed to conceptualizing the client and would have been lost if the reason for the question had not been explored.

In our work with clients, we have had multiple instances where clients have inquired about our sexual orientation. There is some debate among therapists of different orientations and positioning themselves differently along the dimensions we have discussed, about whether LGBTQ therapists *should* be "out" about their sexual orientation. We have spoken with some therapists, focusing more on the social structural context, who take the stance that LGBTQ therapists who chose to not disclose their sexual orientation are contributing to the marginalization, shaming, and heterosexuality or homophobia that is damaging to LGBTQ people. Others, focusing more on an individual level, take the stance that a therapist's orientation, whether LGBTQ or straight, is a personal issue and that such personal issues should not be disclosed. Satterly (2004) presents an excellent model that explores different factors influencing a therapist's consideration of disclosure of sexual orientation, including issues of professional identity and context, and theoretical orientation, with consideration of issues of homophobia, oppression, and authenticity.

In addition, disclosing something from the very beginning of therapy (for example, stating in one's biographical brochure or website that one is "a lesbian therapist specializing in CBT with adolescents and adults" or that one is a "heterosexual therapist who identifies as a strong LGBTQ ally") can be very different than needing to make a decision about disclosure during the course of therapy. In the latter, the meaning may be more determined by the particular relationship with the client rather than the therapist's general intentions. A question from a client during the course of therapy (and the decision about whether to disclose) may have vastly different meanings for different clients. For example, Ellen was a woman in her 20s about to become a graduate student. She was successful in her education and her job and had had several heterosexual relationships of varying lengths and intimacy. She came to therapy in search of self-improvement, to address general feelings of minor anxiety, and to improve her relationships. The therapist saw her for about 14 months and the relationship aspects of therapy reflected, primarily, the realistic relationship, while simultaneously exploring how her past family relationships affected her current relationship patterns. About halfway through the therapy, Ellen asked the therapist about her (the therapist's) sexual orientation. In this case, the question about the therapist's sexual orientation was likely related to Ellen's growing interest in exploring bisexuality and her view of the therapist as a potential model.

In contrast, remember Lindsey from Chapter 4? Lindsey was a lesbian-identified woman in her 30s who was in an abusive relationship. As discussed in Chapter 4, when the therapist inquired about the meaning behind Lindsey's questioning the therapist's sexual orientation, Lindsey responded by saying that if the therapist were a lesbian, then she and Lindsey could have a sexual relationship. The therapist had anticipated that there might be some major feelings behind the question. In addition to the issues related to the abusive relationship, Lindsey struggled with overwhelming debilitating anxiety that initially kept her from working or being able to complete classes in school. She also had a history of physical, emotional, and sexual abuse. The therapeutic relationship with Lindsey had been characterized by moments of intense transference, only some of which Lindsey was consciously aware of. Lindsey's answer about why she was asking about the therapist's sexual orientation did not lead to the therapist's disclosure, but instead to several sessions setting the ethical boundaries and also exploring Lindsey's wish and need to be cared for in ways that were not damaging as in her relationships with others outside of therapy.

Therapists are also faced with dilemmas about disclosure even when a client does not ask a direct question. There may be instances where you may feel that it would be beneficial for a client to have a model of acceptance; to know that a particular experience he or she is struggling with is not odd, unusual, or pathological; to know that the therapist will not negatively judge; to know about options or resources available that relate to your personal experiences; or to know that someone he or she respects (you, as the therapist) has had some particular experience. One of us had been working with an anxiety-ridden adolescent male for three years whom she thought might be gay. About a month before he was to leave for college, he came out to the therapist with a great deal of shame and ignorance about being gay (he thought it was because of his parent's divorce). Because we lived in the same community and his family knew several people who knew the therapist, she asked him if he knew she had a gay son who is married (to a man) with children. It turns out he did not know. The therapist later reflected on her spontaneous disclosure, discussing it with two of her peer supervisors. She was concerned about whether it was therapeutic or whether it was driven by her own countertransference and desire to protect the young man from his parents' rejection. It turned out that the therapist's disclosure enabled the client to ask questions and open his mind to the possibility that being gay was not "the end of the world."

EXERCISE 5.8 ▪ Consider what circumstances, kinds of clients, or client conceptualizations might influence you to feel more or less comfortable disclosing different experiences or statuses listed in Exercise 5.7.

When considering issues and dilemmas related to disclosure, we also want to address ways in which therapists may be disclosing information about themselves without directly verbalizing this information or, sometimes, without being aware that they are disclosing information. A therapist's overall presentation communicates information to a client, just as a client's presentation communicates information to the therapist (information that we have discussed as important to conceptualization!). The style of clothes that a therapist wears, the kind of jewelry (or even the lack of jewelry), whether the therapist wears a wedding or engagement ring, and the décor of a therapist's office all communicate information. This is true even when we try to be "neutral" because the idea of "neutral" is cultured. For example, we know therapists who have French impressionist prints in their office and think of them as neutral and soothing. But an African American, Chinese American, or Mexican American therapist may not make the choice of these European artists as neutral and soothing. And clients from different backgrounds may read meaning into the choices these therapists made, and the kinds of art or artists they chose not to include in their office décor.

In this age of the Internet, another area to consider in relation to disclosure is online social networking and online images. Increasingly, graduate training programs in psychology and mental health are developing explicit policies for trainees about their self-presentations on social networking sites such as Facebook, MySpace, and websites in general. While there is much debate about the balance of graduate training policies and trainees' rights to free speech and representation, the inevitable truth is that any online presence that is accessible by the public is accessible to clients and may, therefore, affect the therapeutic relationship and the progress of therapy. Clients *do* "Google" their therapists; they also do image searches on their therapists. We have heard of instances where clients have found online pictures of their therapists drinking, partying, partially dressed, and kissing or fondling. Therapists do not always consider that their clients might be viewing these kinds of pictures. Even less provocative pictures, such as vacation pictures with family, friends, or partners, may have meaning to clients that affect the therapy.

For example, if Lindsey's therapist had vacation pictures taken with her partner available on the Internet, Lindsey may never have openly asked the question about her sexual orientation, but simply found her answer on the Web. The therapist may have been unaware of the meaning that Lindsey was attributing to the pictures and, therefore, to the relationship with the therapist. Thus, therapists need to make conscious and considered decisions about disclosure in relation to their Internet presence and become familiar with using the privacy settings on social networking sites. Furthermore, because social networking enables other people to share information about you, we encourage trainees to periodically Google themselves and do image searches on themselves. It is important to know what information about you is available to your clients.

Boundary Issues Related to Managed Care Systems

Working in managed care systems can bring up particular issues related to boundaries because some of the structural issues involved in therapy (services provided, ways of addressing issues related to payment, confidentiality of records) may be affected or determined by the system, rather than by the client's own needs or a negotiation between the therapist and client based on conceptualization and treatment planning. This may also be true more generally for therapists working in organizations with pre-set policies and procedures. For example, even if the therapist is able to provide the services needed and desired by a client, these services may not be covered by the client's health care plan. It is incumbent upon the clinician and the client to check to see that services are covered by the client's plan prior to the first session so that there are no surprises. If the plan or agency has polices about cancellations or missed appointments, this also should be discussed at the initial session; for example, if clients will be billed for last minute cancellations or missed appointments, or if clients will be terminated after a certain number of missed sessions. Even if these matters and necessary paperwork have been explained in reception, the clinician needs to ensure that the client is aware of and clearly understands cancellation policies and all aspects of the fee policies.

Another dilemma regarding fees is non-payment. This may be related to therapeutic issues, such as when it may be a form of "acting-out" resistance or expressing anger at the therapist. But it may also be related to external issues of resources, for example, if a client is laid off and loses health insurance or if the benefits are reduced. This is an example of why making the boundaries between the therapy and the organizational context is so important. If a therapist has not done so and has to tell the client that he/she will be terminated due to non-payment unless he/she can develop an alternative plan, just when the client most needs help related to the stress of being laid off, then the client may feel abandoned, angry, and betrayed. If, on the other hand, organizational or managed care polices have been previously reviewed, then it is easier for the client and the therapist to maintain the alliance in addressing the issues. For example, some settings will carry a client for a while or reduce fees, and we have known cases where clinicians have actively advocated with a public health resource to obtain funding for continued treatment. In sum, reviewing managed care or organizational policies is important because the client needs to understand that there may be actions that are not about the therapeutic relationship or the therapist's feelings about the client (or vice versa). Maintaining good boundaries needs to occur not only between the client and the therapist, but also between the therapy and the system in which it is occurring.

A particularly complex issue for therapists dealing with managed care is that of confidentiality. Many managed care plans require detailed information for further authorizations, and they do have access to audit your records. In these instances, it is important to consider whether there is any required information that could prove harmful to the client. If so, it is important to discuss the reporting requirements with the client and to agree on what you must reveal and what is not necessary to reveal. For example, there are some clinicians who do not usually share with the client what they record as a diagnosis because doing so is not part of their theoretical orientation. There are others who find it therapeutic to discuss possible diagnostic categories with clients. The context of managed care may affect these interactions. Some savvy clients want a more severe diagnosis recorded in order to receive more services within a managed care plan. Others may request that substance abuse issues not be part of the record. These are delicate matters that must be considered and decided upon considering institutional policy, managed care policy, legal and ethical issues, ethical stances of the therapist, and best interests of the client.

In recent years, with mental health parity and other health care policies, most managed care systems have demonstrated more sensitivity to these issues and no longer push clinicians to reveal the intimate details of a client's treatment. They focus more on goals and objectives and the therapeutic interventions utilized to achieve them. If a clinician or client runs into difficulties with his/her insurance company, assertive dialogue with supervisors and then management usually brings about an acceptable resolution.

SOCIAL JUSTICE AND CULTURAL SENSITIVITY

Cultural sensitivity is also an ethical imperative. And a part of cultural sensitivity is ensuring that the services you provide and the ways in which you engage in your role as a counselor or psychologist do not contribute to experiences of oppression. But sometimes it is difficult to know what this means. For example, some cultures or subcultures accept or endorse much more severe physical punishment for children than is usually seen as acceptable in the United States. Does cultural sensitivity mean that one should accept or endorse such punishment when working with a client from such a (sub)culture? Within the United States, there are also cultural values that are oppressive. For example, discrimination on the basis of sexual orientation is accepted and legally endorsed in many states: but that does not mean that we, as therapists, should aim to help GLBT people accommodate to second class status. Another example might be a client from a community or family that believes in White supremacy: is it healthy for the individual or the society to leave that belief unexamined or unchallenged?

Our belief is that just because it is cultural does not make it right, healthy, or liberating. What is difficult is determining what the boundary is and when and why it is justifiable to use the power of your role as a therapist to endorse or challenge a cultural practice or belief, especially one that seems related to the client's problems. This dilemma highlights the fact that, while we strive to avoid imposing our values and worldviews on clients, we always inevitably do, at least to some extent. Although we make active decisions to bracket, or hold apart, our own values and worldviews while we explore those of the client, ultimately our collaborative decisions with clients about what is a healthy outcome is informed by our values as well as their values.

Exploring the ramifications of different approaches is the basis of confronting this difficulty. One of us treated a client in her 20s who was suffering from major depressive disorder. She was from a strict Roman Catholic background. She had had sex with her boyfriend and her mother found birth control in her room. Her mother's view was that the depression was punishment from God due to the sin of pre-marital sex. This was, of course, a cultured belief and one that the therapist did not share. The client was struggling with whether or not she, herself, believed this. With this client, it was important for the therapist to understand the role of Roman Catholic faith, practice, doctrine, and culture in the client's life and in her relations with her mother and family. It was possible that the client believed as her mother did or that she did not, but was still struggling with her mother's opinion and blame.

For now, let's assume that the client at least partly believed that her mother was right. One option was to explore diverse ways or approaches to being Roman Catholic, which means, of course, that the therapist must be familiar with these. This approach may help the client negotiate dissonance between her actions (having pre-martial sex) and her beliefs (pre-marital sex is a sin that will be punished). It is an attractive option because it focuses on exploration and enabling the client to see multiple possibilities and make choices among them. However, by exploring multiple meanings of being Roman Catholic, the therapy still presents alternatives to the client's family. If the client were from a collectivistic culture, this might create a new or alternative dilemma: for the client to choose between developing a Roman Catholic identity that reconciles her actions but creates distance with her family, or to endorse her family views. This does not mean the therapist should not take this approach, but the therapist needs to consider the ways that culture interweaves with the multiple layers of context for clients (individual, family, community) and to recognize that choices related to cultural issues will have rippling effects on the fabric of clients' lives, not only on the level of the individual level or the presenting problem but also on the level of the family, community, and other social networks.

EXERCISE 5.9 ▪ Throughout this book, we have presented several cases that relate to the issue of differing cultural worldviews. These cases all raise the dilemma of how the therapist can contribute to positive change without simply imposing values. Review again the following case examples and consider your cultural values in relation to those of the client: John and Jan from Chapter 2 (page 26); Val and Maria from Chapter 4 (page 78 and 80, respectively). What kinds of decisions would you make about endorsing, respecting, challenging, or exploring your own or the clients' values and beliefs about what is healthy? What kinds of ramifications in varying contexts will these decisions have for the client? Try to consider possible negative effects as well as positive effects of different approaches.

Clients' Internalized Oppression

Thus far, we have been discussing dilemmas related to differences between the cultural beliefs of clients and therapists and the need for therapists not to contribute to oppressive experiences, either through imposing values that denigrate or dismiss the clients' cultural values *or* by unquestioningly endorsing or supporting clients' cultural values that may be oppressive or detrimental to their health. An additional dilemma that therapists may face relates to when clients' have internalized oppressive beliefs. For example, when a racial minority client have internalized racism or when a LGB client has internalized homophobia. In some cases, internalized oppression may be central to the presenting problem and, therefore, the associated conceptualization—the case of Peter above may be such an example. In these instances, it is still difficult to know how to approach exploring these feelings, but the question of whether to explore is much more straightforward. In contrast, there are times when clients may present with issues that do not, on the surface or to them, seem related to internalized oppression, but the therapist sees that internalized oppression is affecting the client's well being and psychological health. In this case, questions arise for the therapist about how to integrate these issues into conceptualization and treatment planning.

For example, a Black client may present with test anxiety, which the therapist treats with cognitive–behavioral techniques. However, as the therapy progresses, the therapist begins to hypothesize that part of the test anxiety is related to internalized oppression and the many messages that the client has received throughout his schooling and socialization that he cannot or should not succeed academically because he is Black. The therapist believes that the client is being affected by stereotype threat (Steele, 1997). Furthermore, the therapist sees possible patterns that suggest that the internalized racism is affecting the client's career advancement, relations with peers, and self-esteem.

However, the client has never explicitly brought up being Black as related to the test anxiety, is making good use of the cognitive–behavioral techniques, and feels that the issues that brought him into therapy are close to being resolved.

EXERCISE 5.10 ▪ What should the therapist do? What if the client chooses to continue in therapy, requesting a shift to a less structured approach to address general issues of anxiety, but still does not relate his experiences to his race? What if he were to reject the connection when it was offered by the therapist? Integrating these issues into the conceptualization is imperative, because it helps the therapist understand what might be some of the barriers to change. But if the client does not see these issues as relevant, then such issues may need to be indirectly, rather than directly, integrated into treatment planning and implementation.

EXERCISE 5.11 ▪ A client's internalized oppression may also interact with the therapist's own self-concept or self-reflection process. What might be some of the different issues for a White therapist, a Black therapist, or a non-Black therapist of color in working with the Black client described above? Consider the therapist's own racial identity and relation to possible internalized racism. What if the therapist was White and had never discussed race? What if the therapist was Black and endorsing "immersion" attitudes in racial identity (Helms, 1995)? What if the therapist was Asian American and endorsed the model minority myth?

A characteristic of internalized oppression is that the person affected is frequently not aware of it. In fact, he or she may actively resist the idea that their social status is related to psychological or social differences. People in oppressed statuses may deny the effect of their difference or relative lack of privilege as a way to defend against feelings of powerlessness, pain, or anger (Pinderhughes, 1989). But the experience or effects of internalized oppression may be evident to the therapist, who is trained to see multiple and possible contributors to psychological challenges. Sometimes, the therapist can get caught up in the emotional loadedness of structural issues like racism or homophobia. And sometimes, the therapist can be uncomfortable him or herself with considering these issues.

However, consider the possibility that in some ways, the example above is not that different than any other reason why a client would come to therapy. Clients come to therapy because they want help in changing. This need relates to either not understanding what is going on and thus not being able to act to change it, or understanding and encountering barriers to change which are also frequently not known. The client is seeking the expert knowledge of the therapist, and the skills in working with the client to translate that knowledge into change for the client. Even client-centered therapists, who firmly believe that the knowledge is in the client not the therapist, believe that they are offering something for their clients.

We have discussed how the therapist may conceptualize the influence of attachment, relationship patterns, and reinforcement experiences as contributors to presenting problems. And, frequently, the client may not be fully aware of these influences or the connections between them. Internalized oppression is a particular conceptualization of a presenting problem, attending to the structural as well as family and personal levels. As with any conceptualization, the extent to which you share it with your client and the extent to which you focus on what you see as the underlying issue related to your conceptualization will vary depending on your orientation and on the client's particular position, goals, and readiness to address different aspects of their issues.

Clients' Bias and Discrimination Related to Therapist Minority Status

A final issue that we want to discuss (while acknowledging that there are many others we are not discussing due to time and space constraints) in relation to social justice issues is when a client who is in the dominant status in relation to the therapist is either indirectly or directly endorsing discriminatory attitudes that relate to the therapist. Frequently, these issues are brought up somewhat indirectly, and the therapist may need to make decisions about whether or how to address them. Examples include: (a) a client who says to an Asian American therapist: "I am surprised you speak English so well!"; (b) a White client describing her experience of being stopped by a policeman for speeding to a Black male therapist who says "Oh, you know how the police are when they know and you know you've done something wrong!"; (c) a client who asks a Mexican American therapist with an accent where they are from and how long they have been in the United States; or (d) a client who looks surprised when the therapist identifies herself as Native American and says "I thought the Indians were all killed."

EXERCISE 5.12 ▪ Consider the examples given above. If you were the therapist described, how would you feel in the moment? How might these interactions affect your feelings about the client and the relationship? How would your thoughts and feelings be affected by other variables, such as the developmental stage of the therapy, that is, in what session does this occur? The client's presenting problem? What might be some of the meanings behind these questions or statements? Are they all definitely examples of discrimination? How would you respond to the client?

In instances like these, the therapist will need to consider how various responses (including not responding at all) will affect the therapy and the relationship. We believe that statements like those expressed by the clients above are important and should not be simply ignored, while we also recognize that it can be emotionally loaded for a therapist from a minority status to address these issues. We recommend first working with the client to clarify their meaning. Some questions to consider as you inquire about the meaning behind the statement might be: Does the client have concerns that the therapist may not be able to understand their experiences? Does the client have active stereotypes that are affecting their ability to utilize the therapist's expertise? Are the client's biases directly related to the presenting problem (perhaps the client is struggling with relating well to work colleagues that are Black), indirectly related (perhaps the client has difficulty seeing people for who they are rather than who they expect them to be, both good and bad), or seemingly unrelated (perhaps the client is struggling with an abusive relationship with his/her partner who is of the same race and ethnicity as the client).

Sometimes, these attitudes might become directly evident, as in the example of the White Vietnam War veteran working with the Asian American therapist. In these cases, it is relatively straightforward in the moment, as it becomes part of the therapy discussion. But even if it is clear to the therapist that it must be addressed in the moment, it may elicit feelings within the therapist that will need to be processed in consultation or supervision. We recommend addressing these kinds of incidents both in the therapy and within the therapist because we believe that the clients *are* communicating something relevant and that therapists *are* affected by their clients. To not address these moments means that we cannot understand them and work through them to ensure that they do not damage either the client or the therapist. However, there is no simple answer to how to address these moments of "microaggressions" (Sue, 2010).

CONSULTATION AND SUPERVISION

In addressing any dilemma in therapy, one of the best strategies is seeking consultation and supervision. Consultation, supervision, and, sometimes, specialized training is particularly important as you venture into new areas of work. But consultation is always a good idea when facing a dilemma, no matter how experienced one is. This could be a quick telephone call with a peer or a more extensive consultation with a paid consultant or expert. Consultation and supervision are tools to help therapists re-consider and, at times, re-conceptualize. They enable us to step back and articulate the dilemmas, considering multiple influences and possible outcomes. They encourage us to examine our own biases and preferences by providing an alternative view. They also enable a "reality check" of what is ethical or acceptable in the field.

On-going peer supervision or consultation is desirable: so we do not get stuck in conceptualization and intervention ruts; so we are helped to become aware of what we may have missed; so we may continue to look at people and their circumstances from multiple perspectives; so we may work through our own reactions and feelings; and so we can name and identify our problems *and* our progress in order to facilitate our ongoing growth and expertise as therapists. Therapy can be very isolating; since ethical codes preclude us discussing our clients, except with supervisors and consultants, it is important that we arrange for such invaluable feedback.

SUMMARY

We have presented in this chapter some of the typical clinical dilemmas that can occur in therapy. While we have mentioned some ethical issues, we expect that you will have a deeper study of this kind of material in required ethics courses. It would be nice if there were definitive answers for every situation, but the truth is that many complex dilemmas concern gray areas, where there is not a "right" or "wrong" answer. For example, some of the issues of confidentiality and dual relationships are clearly spelled out in the mental health professions' codes of ethics with the exceptions listed, but the actual details that may emerge may not be as clear-cut. Therapist disclosure and boundary issues such as attending a client's wedding and giving and accepting token gifts need to be considered case-by-case in multiple contexts. Your theoretical orientation, dimensional preferences, conceptualization, and clinical skill in assessing what is best for the client and the therapeutic relationship at a given time will help you consider all aspects of these dilemmas.

REFERENCES

Anton, B.S. (2010). Proceedings of the American Psychological Association for the legislative year 2009: Minutes of the annual meeting of the Council of Representatives and minutes of the meetings of the Board of Directors. *American Psychologist*, 65, 385–475. doi:10.1037/a0019553.

Barlow, D.G. Allen, L.B., & Basden, S.L. (2007). Psychological treatments for panic disorders, phobias, and generalized anxiety disorder. In *A guide to treatments that work*. (3rd ed.) P.E. Nathan & J.M. Gorman (Eds.), pp. 351–394. New York, NY: Oxford University Press.

Chang, D.F. & Berk, A. (2009). Making cross-racial therapy work: A phenomenological study of clients' experiences of cross-racial therapy. *Journal of Counseling Psychology*, 56, 521–536.

Ferguson, S.A. & King, T.C. (1997). There but for the grace of God: Two Black women therapists explore privilege. *Women and Therapy*, 20, 5-14.

Helms, J.E. (1995). An update on Helms's White and People of Color (POC) racial identity models. In J.G. Ponterotto, J.M. Casas, L.A. Suzuki, & C.M. Alexander (Eds.), *Handbook of multicultural counseling* (pp. 181–198). Thousand Oaks, CA: Sage.

Karlsson, R. (2005). Ethnic matching between therapist and patient in psychotherapy: An overview of findings, together with methodological and conceptual issues. *Cultural Diversity and Ethnic Minority Psychology*, 11, 113–129.

Okun, B.F. (1990). *Seeking connections in psychotherapy*. San Francisco: Jossey-Bass.

Pinderhughes, E. (1989). *Understanding race, ethnicity, and power: The key to efficacy in clinical practice*. New York: Free Press.

Pope, K.S. & Vasquez, M.J.T. (2010). *Ethics in psychotherapy and counseling: A practical guide* (4th ed.). San Francisco: Jossey-Bass.

Satterly, B.A. (2004). The intention and reflection model: Gay male therapist self-disclosure and identity management. *Journal of Gay & Lesbian Social Services*, 17(4), 69–86.

Steele, C.M. (1997). A threat in the air: How stereotypes shape the intellectual identities and performance of women and African-Americans. *American Psychologist*, 52, 613–629.

Sue, D. (2010). *Microaggressions in everyday life: Race, gender, and sexual orientation*. Hoboken, NJ: John Wiley and Sons.

Whittal, M.L. & O'Neill, M.L. (2003). Cognitive and behavioral methods for obsessive-compulsive disorder. *Brief Treatment and Crisis Intervention*, 3, 201–216.

References related to APA and Guantanamo Bay:

http://www.apa.org/news/press/statements/interrogations.aspx

http://kspope.com/interrogation/home.php

http://www.apa.org/news/press/statements/work-settings.aspx

http://www.apa.org/news/press/releases/2008/09/detainee-petition.aspx

http://www.apa.org/monitor/2008/11/interrogations.aspx

http://www.apa.org/ethics/code/index.aspx?item=15

http://www.apa.org/about/governance/council/policy/sexual-orientation.aspx

6

Beginning Conceptualization: Gathering and Integrating Information

Your theoretical orientation is about people in general. In the first four chapters, we explored your worldview, beliefs about people, understandings of health and pathology, and exploration of contexts. We also introduced dimensions of context and relationships that can help you develop an integrative theoretical orientation and understanding of change processes in therapy that matches your beliefs and style. But your case conceptualization is not a cookbook application of your theoretical orientation. It is a tailored and specialized application of your theoretical orientation, an integration of your worldview and beliefs about people and the process of change with specific information about a client in his or her unique contexts. Thus, we turn now to the process of gathering, organizing, and integrating the information about clients that is the foundation of client centered case conceptualization.

BEGINNING THE PROCESS OF CASE CONCEPTUALIZATION: GATHERING INFORMATION

Information for case conceptualization will come from a variety of sources, such as things the client says and does, our reactions, our knowledge about different contexts and mental health, information from other sources, and so forth. Each piece of information is like the piece of a puzzle. As therapists, we work to decode and assess the meaning of the information and then fit the possible meanings together to try and see the picture as a whole using our theoretical orientation as a guideline. The picture that begins to emerge helps us to generate hypotheses about what is going on overall for the client and how best to intervene. In addition, in the process of integrating and relating the pieces, we identify areas that we need to know more about, the pieces we are currently missing that may be important to seeing the overall picture. Our hypotheses and further explorations then feed back into expanding and clarifying the puzzle picture (conceptualization) and identifying appropriate interventions. The pieces of information we gather may come directly from the client in multiple ways (primary sources) or may come from other people (secondary sources).

Our first contact with or about a client immediately begins providing information that can contribute to our case conceptualization. The *referral context* frequently provides at least some information about the client, his or her contexts, and his or her attitudes toward therapy. How does a client come to you? Is it self-initiated or did someone such as a loved one, a teacher, or a doctor "send" or encourage them to seek therapy? Is the referral voluntary? Is the referral involuntary? Did they select a therapist or were they directed or assigned to you? Do they have private or public health insurance or will they self-pay? Often, we may also learn about the client's presenting problem from the referral source (a secondary source of information) or during the initial contact with the client to set up an appointment (a primary source).

When we first meet a prospective client, we immediately become aware of the client's overt characteristics, such as age, gender, race, general appearance, and style of presentation. While we will eventually need to ask ourselves how the client's gender, race, and other characteristics are affecting his or her presenting issues and general functioning, at this point we need to be careful about not making assumptions based on these characteristics. And we need to be open to being wrong: a multiracial person may appear to us to be racially White, but actually identify as Native American (or vice versa), or a person who appears to us as a woman may identify as a man. In general, we want to begin our information gathering as openly as possible so that we can understand it from the client's perspective and in the client's particular contexts.

From the moment we first learn anything about a client, we begin to generate hypotheses about the client's functioning, about the central problems that are bringing the client to therapy, about how these problems developed and are maintained, and about how we might help the client to change. Depending on our orientation, we may attend more or less to hypotheses about problems, about development, or about strengths and resources. Narrative therapists, for example, may actively work to not conceptualize problems specifically, but rather to understand the overall story that reflects the complexity of the client's experience. Regardless of what aspect of the client our hypotheses focus upon, these hypotheses need to be held very tentatively, especially in the beginning. Part of the goal of gathering information is to "test" these hypotheses, to determine whether they apply to this particular person in this particular context, and to modify them using the new information we are gathering. We need to have confidence that our understanding of the client is sufficiently thorough and specific to the client's unique circumstances in order to identify the best interventions. This does not mean that we need to know everything about the client, but simply that our knowledge about the client should be guiding our choice of interventions.

Interventions in the first few sessions of therapy will be more "generic" and applicable to many clients in many contexts. These kinds of interventions require less specific knowledge about the client in context. The experience of a good working alliance, a caring listener, and the reserved time to explore experiences and problem solve are all helpful, and generic, interventions. But most clients will need more than these generic interventions. In order to choose the most effective interventions, we will need to understand the client more thoroughly. This is a continuous process: therapy usually becomes more complex and tailored to a client's particular experience as time goes on.

Primary Sources of Information

In the first contact with a client, therapists generally begin the conceptualization process by asking clients about themselves and observing them. Therapists also reflect on their responses to the client and the client's responses to them. These three sources of information (the client's direct expressions, observations of the client, and observations/reflections about the relationship) are continuous sources of new information throughout our interactions with the client and, therefore, contribute not only to our initial case conceptualization, but also to our continuous reconsiderations. We may also choose to include sources of information outside of our direct interactions with the client, such as information from the client's family or significant others, consultations with other providers, or formal assessments.

Information from the Client's Direct Expressions The primary source of information with most adult clients in most types of psychotherapy is what clients actually tell you and how they relate this information. There are, of course, instances where there are obvious constraints on the information that clients are able to offer. For example, clients with autism, psychosis, dementia, or other cognitive or language impairments may not be able to accurately or coherently articulate information about their own experiences. In these instances, other sources of information (observations, collateral information, assessments) become more primary. However, the client's direct expressions are still of great importance, because what they cannot tell you and the ways in which their verbalizations are incomplete, incoherent, inaccurate, or otherwise different from others are a central part of the information you will use for case conceptualization. Even when clients seem highly articulate and insightful, the information that they do not tell you, what they avoid discussing or have difficulty articulating, can communicate much about their experiences and struggles.

There may also be other issues, not related to pathology or disability, that affect the kind or quality of information that clients are able to offer. Clients who are not able to write in English (either because they are not literate or because they are literate in a language other than English) may not be able to complete a written intake form or assessment in English. Clients with limited English proficiency or even relative verbal fluency may also have difficulty with written intake forms or assessments; they may not understand the vocabulary used in these forms or their English reading and writing may not be as fluent or expressive as their spoken English. Bilingual clients with limited English proficiency may also be less expressive or complex in their expressiveness if the therapist only speaks English and the therapy is limited to this language. This does not mean that they do not think or understand in complex ways, but that the language demands and limitations of the therapist (for example, that the therapist is not capable of speaking their language) limit the clients' ability to communicate their complex understandings. Working with a client who is monolingual in a language different from that of the therapist has additional challenges related to the need for interpretation. There are multiple issues raised by interpretation in relation to what is or can be expressed fully and the ways that the presence of an interpreter changes the therapy relationship and communication (for examples, see Tribe, 2007; Miller et al., 2005).

Language issues may also affect bilingual clients who are highly proficient in English; we have heard of a case where a bilingual therapist was working with a bilingual client primarily in English. The client's fluency in English and in Spanish was comparable. After many sessions, they started speaking in Spanish and, for the first time, the client spoke extensively and emotionally about her relationship with her mother, with whom she had spoken only Spanish. The therapist and client came to understand that speaking in English had inadvertently led the client to avoid certain areas and topics that she thought about in Spanish.

In sum, it is important to consider how the client's presenting problems and cognitive capabilities may affect the kinds of information available and the best ways to obtain information. It is also important to consider how the client's and your own language proficiencies will affect the communication and the information obtained.

EXERCISE 6.1 ▪ In triads, take turns role playing a counselor/client/observer at the beginning of the first therapy session. What kinds of questions might you ask as a counselor? How would you ask them? Remember, your initial goal is to learn about the client, why and how they have come to therapy, and how they understand their presenting concerns. When each of you has completed the counselor role, discuss what were the most helpful questions for meeting these goals.

How you elicit information will be affected by your theoretical orientation, your worldview, and your preferences, as discussed in earlier chapters. For example, if you are more influenced by a psychodynamic or client-centered orientation and have a non-directive style, you may smile slightly and wait for the client to begin. Or you may ask some general questions or start with a lead such as "How did you decide to come for therapy?" "How do you think I may help you?" or "Tell me about yourself," but will basically let the client take the lead in revealing him- or herself and issues. If you are focused on cognitions and behaviors and have a directive style, you may ask direct questions about the behaviors and thoughts that are troubling to the client, such as "What are your biggest concerns at this time?" "What would you like to have happen here?" A constructivist approach may start with a lead such as "Tell me your story." The types of questions that clinicians might ask are usually open-ended in that the client cannot answer with one or two words. The purpose is to engage the client in the process of articulating his or her understanding of the problem(s).

The actual information you receive will also be affected by your theoretical orientation and dimensional preferences. If you conceptualize problems as related to early development, you will seek out more information about the client's early experiences and relationships than a therapist who is more present focused. If you conceptualize problems as related to structural issues, you will seek out more information about the client's social status identities than a therapist who is more individually focused. Your preferences will affect your relative emphasis, but it is important to have some understanding of most areas and contexts for the client, even if they are not your preferred point of intervention.

There is a multitude of ways to gather information from clients. The simplest is to ask the client about her/his experiences, either in writing (for example, an intake form) or verbally in the session. But there are many other ways to structure the process of obtaining information from clients directly. One might encourage the client to use metaphors and analogies, complete a time line, construct a genogram together, use drawing or journaling or other expressive techniques or engage in role-playing or psychodrama. For example, with Nancy, verbal communication and observation of associated nonverbal presentation was effective. With another client who was more visual and preferred more structure, the clinician used a pie chart to elicit different contexts and experiences within each context, such as the size of wedge for family, work, leisure, peer relationships, extended family, and other salient areas. There are a multitude of resources that describe different ways of obtaining information from individual clients and families in early sessions. Our purpose here is not to review these details, but simply to encourage you to be aware that you are actively making choices about how to obtain information, and these choices will affect what information is offered by the client. The choices you make relate to your dimensional preferences and related theoretical orientation.

EXERCISE 6.2 ▪ Make a list of all the ways you know to gather information about the client's presenting problem and contexts. What strategies or approaches are you most comfortable with? What strategies are you least comfortable with? What determines your approach to gathering information? What are three ways of gathering information that you have heard or read about that you would like to learn? Why do these approaches appeal to you? In what instances would you be most likely to use them? How do your choices relate to your dimensional preferences, that is, your understandings about how people develop and change, your emphases on different client contexts, your preferences about therapist style, or your reflections on relationship dimensions discussed in the previous three chapters?

Usually, information from the client initially focuses on the client's presenting problem and on understanding the client's experience of or relation to his or her multiple contexts. These areas include not only what the problem or context experience is, but also how the client feels about the experiences, and how he or she responds to the experiences and the feelings. Because most therapists do focus on the presenting problem, we frame our discussion of gathering information in relation to this, but the discussion of sources of information is more general—for example, one could gather information from various sources about the client's strengths or overall story of their life or development, rather than about the presenting problem.

Information from the Client About the Presenting Problem. What the client tells us about his or her problems represents the client's "story" or "narrative" of perceptions, experience, and interpretation. Here, we seek to understand what brought the client to therapy at this time, which areas of their life are more difficult than others, as well as which areas are strengths or resources. It is important to remember that these reports are the client's perceptions and not necessarily "facts." For example, other family members, or teachers, employers, or peers may have different perceptions and interpretations.

Let's consider some case examples:

Steve, age 35, requested an appointment for "depression." On the telephone, he told the therapist that he was no longer able to tolerate the pressures at his computer jobsite where layoffs were fairly routine and then coming home to a wife and three young children who were "noisy" and interfered with his need for relaxation. His physician had recently prescribed medication, which "wasn't working." In short, he reported feeling "overwhelmed" by life's responsibilities and he felt overburdened by being "trapped at such an early age."

Marie, age 30, was referred for biofeedback and relaxation training due to wrist and arm pain and related anxiety from work-related repetitive movement injury at her factory job. She described well the repetitive movement resulting in the injury, the type of pain, and the movements or actions that caused pain or did not cause pain. She also described being anxious and fearful about several related issues: whether she would be able to fully recover and return to work; whether she and her husband could support themselves and their two children if she could not return to work; and whether her work performance would be negatively evaluated because she had filed the workers' compensation claim.

Yuki, age 23, was referred to a college counseling center by the college health center due to complaints of headaches, racing heart and sweats, and decreased interest in her school work. She described her symptoms as beginning "two weeks after this thing with this guy I went out with," a description she had not offered to the staff at the health center. Yuki initially expressed being most concerned about how her physical problems were keeping her from regularly attending classes or concentrating on her work and her slipping grades. As an international student from Japan in her first year of graduate school, she also attributed her difficulties to being homesick and described feeling angry with herself that she could not better manage the transition to the United States and graduate school.

Although clients' initial descriptions of their presenting problem often relate more to their own styles, self-understandings, and understandings of therapy and the therapy relationship, most therapists will want to understand the issues in greater depth than the client first describes. Thus, the therapist will ask questions and explore greater details about the presenting problem. What the therapist attends to and how they ask questions is affected by the therapist's dimensional preferences and related theoretical orientation. We are also continuously trying to decode the messages in order to uncover deeper meanings.

EXERCISE 6.3 ▪ Using the examples of Steve, Marie, and Yuki, what information would you want to elicit in each case? What would your theoretical orientation emphasize? What contexts would you want to explore? Would you be more interested in the present, past, or future? How comfortable would you be working with each one? What might be the relationship issues in each case? Discuss these questions in small groups.

Information from the Client About Client Contexts. In order to fully understand the client's presenting problems, we must understand the client more generally, not just in relation to the presenting problem, which is what we mean by understanding the client's contexts. This includes the client's individuality (such as personality, personal experiences, preferences) as well as various aspects (such as relationships, environmental aspects, ideologies, relationships) of different contexts (such as family, extrafamilial, community, and sociocultural), as discussed in Chapter 3. When clients talk about the kind of person they are or the things they like or dislike, they are discussing their individual context (although inevitably interacting with other contexts). We also frequently want to know about the client's family (past and present) and significant extrafamilial relationships, social groups and roles, community, and social structural identities and contexts. We might explore these contexts as they come up in relation to discussing the presenting problem, ask about these areas directly, or use other structured means to explore them such as a genogram or sharing of a photo album. For each particular client, different contexts may be more or less important, depending on the presenting problem.

In exploring contexts, it is important to attend to the language we use and the assumptions that may be reflected in this language or in other aspects of our questions. If we ask about marriage, for example, there is in most cases an assumption of heterosexuality, as marriage is not permitted for gay and lesbian couples in most states at this time. Another erroneous assumption that is frequently made is that the sex of the current intimate partner is an indicator of the sexual identity of the client: that is, if a female client is in a relationship with another woman, then that client is lesbian. In fact, that client may identify as lesbian, bisexual, heterosexual, or some other way. Another example is related to family background: there is frequently an assumption that the primary caretaker of children is the mother, but fathers, grandparents, and other family or non-family members may be the primary caretakers. A final example is an assumption that the race or culture of a client's caregivers or parents is the same as the race or culture as the client. It can set back the working alliance for the client to have to correct a therapist's assumption, so it is best for therapists to consider carefully how they approach exploring contexts.

EXERCISE 6.4 ▪ In the paragraph above, we identified examples of when a therapist might make or convey an erroneous assumption. What language might you use in asking questions about or discussing these areas in order to avoid communicating assumptions?

Some assumptions are conveyed directly in language and can be avoided by using language more carefully. For example, one could ask a client whether he or she has a partner, rather than whether he or she is married or has a boyfriend or girlfriend. Or one could ask about primary caregivers rather than parents, or preface this discussion by asking "Who took care of you most in your childhood?" Many assumptions are avoided simply by realizing that one needs to ask the client, not make the assumption. If you are gathering information about intimate relationships and a female client says that she has a boyfriend, you can ask whether her previous relationships have also been with boys/men and how she identifies in relation to sexual orientation. Asking these questions avoids making the assumption that she only has heterosexual relationships and that her sexual orientation or identification is related to the gender of the partner with whom she is currently involved. If you have a written intake form with demographic information, you can also integrate your awareness of avoiding assumptions there; for example, in addition to asking about the client's racial, ethnic, and religious identities, you can ask about these identities in relation to the client's family, parents, or primary caregivers. You can also include a wider variety of choices if you have choices at all. For example, you can include "transgender," "transsexual," "queer," and "other" options for gender.

Let's return to our case examples, to examine what kinds of information clients may initially offer about themselves, their problems, and their contexts:

Steve's understanding of his problems was directly related to his family and work contexts, so it was particularly important to immediately understand these contexts. In describing his presenting problems, Steve spontaneously described some aspects of these contexts and others were explored through more direct questions or structure from the therapist. Steve reported that his wife was complaining that he never did enough around the house or with the children and that she was becoming "fed up" with his "self absorption." Steve was bewildered by his wife's complaints. A genogram revealed that Steve was the oldest and only son of three and had been doted on by his mother and sisters. Steve's father was a traveling salesman who was rarely home; when he was home, he was treated royally by his wife and daughters and favored Steve as his only son. With regard to education, Steve had always been a middle-level student who "played" more than he studied. At his current place of employment, Steve's last few performance reviews had been mediocre because his supervisor did not feel Steve was giving his work full attention. Steve felt that the reviews were "unfair."

Technically, it would have been possible to provide Marie with biofeedback and relaxation training to address her pain without exploring her contexts. Biofeedback could also have been helpful in addressing some of the related anxiety from a physiological stance. However, issues such as treatment motivation and compliance with physically related interventions are affected by intrapsychic, interpersonal, and social contexts. In addition, pain and anxiety are not solely physical issues, and psychological interventions can contribute significantly to their alleviation. Marie's contextual picture was much more complex than her description of her initial presenting problem. She described herself as rarely complaining. She had experienced pain in her wrists for some time before seeking a doctor. She experienced her work environment as "hard" and "uncaring" and described how the factory line workers were underpaid, working in poorly vented conditions (a particularly major problem given that the products produced odors), and that the system of quotas for production frequently lead workers to not take breaks and ignore physical stresses because they would not be able to make their quota. Marie came from a poor family whose members sometimes did not have enough to eat. In later sessions, Marie revealed that her husband was physically abusive.

Yuki's family in Japan was close and supportive; she spoke with her mother and sister several times each week. Her family was upper class and her parents were highly educated professionals. Her parents and extended family were proud of her accomplishments and were financially supporting her graduate study in the United States. She had a couple of close friends at school in the United States who were also international students from Japan. As Yuki had related her symptoms to a particular incident, the therapist sought to understand what had happened in her context at this time. As she described "this thing" it was clear that she had been date raped two months prior to seeking counseling, although she did not use these words. Although Yuki did express feeling upset about the date rape, she had told no one and said that she "just needed to get over it." She felt that the rape was her own fault and believed strongly that her friends and family would feel the same.

Understanding clients' contexts enables us to make connections between the "symptoms" and the lived experiences in order to understand the possible meanings and functions of the problems that the client is facing. This helps us to understand how the problems and the clients' understandings of the problems are embedded in clients' lives; how they are connected to various areas of functioning that may be experienced as unproblematic or even necessary; and how having the problems or understanding them in certain ways may actually be seen as helpful by the client, although he or she may not be aware of this. We can then consider what might be challenging about creating change and tailor our choices about interventions with this in mind.

EXERCISE 6.5 ▪ How do you now understand the presenting problems of Steve, Marie, and Yuki now? How does the information about their contexts shape your thoughts and questions about the development or current experience of their presenting problems? What are your preliminary thoughts about how therapy might help them address the presenting problems?

As Steve talked about his work and family, it became increasingly clear that he was feeling burdened and unfairly judged in both areas. Understanding and exploring Steve's family of origin and his expectations that husbands and fathers should be "treated as royalty" led the therapist to wonder whether Steve's expectations about his roles and responsibilities were contributing to his anger and disappointment.

Understanding Marie's contexts helped the therapist understand that there would be a continuing risk for additional physical injury at work and to consider both physiological and psychological interventions to help Marie avoid this in the future. Understanding the financial pressure on Marie, its connection to her early experiences of poverty and hunger, and her commitment to keep her children from having similar experiences contributed to hypotheses about the challenges for Marie in taking care of herself, particularly if this self-care was seen as risking the comfort of others. The therapist thus considered the usefulness of interventions aimed at validating and prioritizing self-care, but framing these in relation to caring for others. However, this approach was significantly complicated by the revelation that Marie's husband was abusive, which became a second primary focus of the therapy.

Yuki's initial presentation of her problems focused on physical symptoms and adjustment to a new cultural context far away from her family. However, even though Yuki dismissed the significance, the therapist understood the interpersonal context of the rape as central to understanding her presenting problems. As Yuki described her friends and family and then described her silence about the rape, the therapist conceptualized her presenting problems not only in relation to a traumatic event, but also in relation to intense isolation in relation to that trauma.

The three cases above emphasize that in addition to sensitivity, we need some knowledge about larger contextual issues such as social class, gender roles and identity, cultural background and ethnicity, sexual orientation, and their specificity for a given client, as discussed in Chapter 3. Clients may not see the impact of these issues because they are embedded in their own experiences. For example, Steve came from a traditional east European Jewish heritage, which valued male children more than female children. For Steve, this value was "normal" and he did not consciously see how this shaped his current family relationships and expectations. Understanding this cultural context helped the therapist to understand Steve's ideas of male privilege and how having three children close together in early marriage took his wife's sole attention away from him, which contributed to his feeling resentful and devalued. Another example is Marie's social class background, which likely contributed to her stoicism at work and her intense fears about financial stability.

EXERCISE 6.6 ▪ Yuki's therapist knew that research conducted with White European American rape survivors indicated that self-blame and depression were frequent responses (Neville & Heppner, 1999). She also knew from her prior experiences with rape survivors that it was frequently helpful to break that silence and that the self-blame and shame experienced by rape survivors could be lessened by positive responses from supportive friends and family. However, Yuki insisted that her Japanese friends and family would completely agree that the rape was Yuki's fault. The therapist did not know how Japanese culture understood rape and rape victims and gender roles. What information does the therapist need to understand the contextual effect on Yuki's problems and what interventions would be best? How might the therapist find out more information about the cultural norms? If a cultural belief is harmful to a client, how should the therapist respond?

Another case involves Lucas, a White European American whose father came from a Midwestern family with the value "spare the rod and spoil the child." Lucas mentioned to a classmate in eighth grade that his father had "caned" him because he had gotten a failing grade on a math test. When the classmate told the teacher, he was obliged to refer Lucas to the school social worker. The school social worker talked to Lucas, who accepted physical punishment as "natural" and did not understand the fuss. In turn, Lucas's parents were upset by the school's "interference" into their family matters. The school social worker, who was required to report this situation, worked in close collaboration with a sensitive social worker from family services to supportively join with the parents and help them to understand why this type of punishment was unacceptable in contemporary U.S. culture. They encouraged the parents to attend a community parents' group and learn effective behavioral management skills. This example emphasizes that it is not only "different" ethnic/cultural/national contexts that influence what is seen as normative.

Remember that verbalizations from the client limit the information garnered to what the client is consciously aware of and able to articulate. As we see with Steve, there are limitations in the client's verbalizations: Steve's lack of awareness and limited viewpoint are what brought him into therapy in the first place. Sometimes, clients come in because someone else has told them what their problem is and the client does not really see that. As a relationship is established, the therapist and client together can agree on what they want to work on but it may not be immediate. Whether or not the counselor feels power and status, the nature of the helping, teaching, or supervisory relationship has inherent power and status. Our job is to be aware of this power and status and not to abuse it. So we must always be aware that clients' perceptions of our power, status, and the treatment context may affect (consciously and unconsciously) what, how, and when they choose to verbalize. At the same time, the therapist does bring expertise and training, which is why the client seeks therapy in the first place. Being aware of one's power in the role of a therapist does *not* mean not using it to benefit the client.

The context of a given therapy varies in relation to the particular relationships, as well as the environmental aspects and ideologies. For example, some therapies take place in environments with policies about the allowed number of sessions, which will influence the nature of the work. If one has limited sessions, there may not be sufficient time to develop the depth of trust in the relationship so deeper problems or concerns can be expressed. As in any contextual consideration, the environmental aspects interact with relational aspects in this and other contexts: if the client has never experienced a trusting relationship *and* there are limited number of sessions, it is particularly unlikely that deeper underlying issues will have an opportunity to surface.

Let's consider some of the issues of the therapy contexts for the three clients we have been discussing: Steve's employer had tried to send him to an Employee Assistance Program (EAP) counselor. Steve did not trust that anything he said to the EAP counselor would really be confidential, so he was intensely guarded during those eight sessions and focused only on superficial topics. In his current therapy, he was still initially guarded, but more open because he had sought out the therapist independently. Marie's sessions were limited by her worker's compensation referral, which also required a primary focus on the pain from her injury. While the therapist and Marie addressed the spousal abuse, this could not become the primary focus of this particular time-limited therapy. Although Yuki did not directly discuss confidentiality, the therapist had experience with other students who expressed concern about confidentiality despite the counselors' insistence that communications were privileged. The therapist wondered if Yuki's initial strong focus on academic issues was at all related to being seen within the college counseling center.

In sum, information from clients' direct expressions about their presenting problem and related contexts is affected not only by the choices and understandings of the clients, but also by the therapists' dimensional preferences and related theoretical orientation, and by the context of the particular therapy. The therapist's dimensional preferences and related theoretical orientation will influence the information that is solicited, as well as the approach to gathering that information. And the choices of both client and therapist about what information to focus upon and how to share information will be influenced by their understanding of the therapy context. Thus, gathering information is not a neutral activity, but is an interactive process actively shaped by the therapist and the therapy context.

EXERCISE 6.7 ■ When you first meet with a client, what information is most important for you to have within the first session? Within the first three sessions?

Within the first five sessions? Why is this information most important to you? How do you get this information? What do you actually say to the client? What information seems *less* important to you? Why might this be so? How might your answers be different with different kinds of clients or presenting problems? How are your answers affected by the context of the therapy? Would it be different if you had only five sessions versus having unlimited sessions?

Clearly, we are gathering a lot of information from clients in these initial sessions. Trainees sometimes question whether they can or should take notes in order to accurately record the information. This will depend on the norms of the organization within which you work, the issues and preferences of each particular client, and your own approach and philosophy. Some theoretical approaches or activities may have forms or written assessments integrated as a matter of course. Other approaches feel strongly that writing notes of any kind during sessions detrimentally affects the connection between the client and the therapist. Some clients will be strongly affected by the therapist taking notes, either because of a relational feeling or approach or because of their specific presenting issues. For example, if the client is presenting with paranoia about being watched or monitored, taking notes is very likely to have a major effect!

We feel that the most important thing to convey to the client in these initial sessions is your interest, respect, and connection to them. So we encourage you to consider how taking notes will affect these issues. Many novice therapists find it challenging to be fully present with the client and to convey that presence while taking notes of any kind and even more challenging while taking more detailed notes. Thus, in most cases, we recommend that you take as few notes as possible in the first session so that you can maintain eye contact and be fully present with the client. If you have particular difficulty remembering information, you can jot down a word or two to remind yourself later. We also recommend that novice therapists build in time immediately after initial sessions to write more detailed process or information notes as they train themselves to remember.

Information from Observations of and Interactions with the Client In addition to information that clients directly provide, therapists gather information from their observations of and interactions with clients. Observation is an important method of eliciting information and frequently guides some of our questions and hypotheses. Observation and decoding verbal messages are assessment tools that cut across theoretical orientations, allowing for different emphases and foci. Therapists observe clients' behaviors, patterns of interacting with the therapist, and their own (therapists') reactions to clients. These are all sources of information for conceptualization.

Information from Observations of the Client. Our observations of clients include their physical appearance and presentation, their movement, their facial expressions, and many other things. As a therapist, you may have different thoughts about a client who complains of depression who seems to bounce on her/his feet, moves and speaks quickly, is exceptionally well dressed and groomed, pumps your hand enthusiastically, makes constant eye contact, has great animation in her/his facial expressions, and smiles frequently compared to a client who complains of depression who moves and speaks very slowly, is disheveled in dress and grooming, limply shakes your hand, and stares off into space while you are speaking.

Observations of *when* and *how* clients respond facially or in their body are also important sources of information. When do clients laugh or smile or frown? Do they avoid eye contact, cross their arms, lean towards you or face away from you at some times and not at others? Thus, it is important to attend to how presentation changes, both momentarily and over days or weeks. The way that you understand the content of a client's communications (information from the client directly) will also be affected by *how* the client communicates this information. You will note whether the client's communications are focused, random, concrete, or abstract. How hard do you have to work to engage the client, to elicit information? Do you have to bring her or him back on track or does the communication flow easily? These kinds of observations affect how you use information from the client in your conceptualization. For example, sometimes what the client is doing and what the client is saying are not congruent. Yuki stated that her headaches and schoolwork were most concerning and shaming to her and that the rape was a minor occurrence. But when talking about her headaches and schoolwork, Yuki made occasional eye contact and leaned toward the therapist. In contrast, when she talked about the rape, she averted eye contact, pulled back, and rounded her shoulders inward.

When clients enter the room, what can you tell from their facial expression—are they sad? Anxious? Angry? Pleased? Is their expression about you, the therapy, something else they are thinking about? Our observations tell us about the clients' thoughts and feelings. However, the ways that people respond vary and are affected by individual style, developmental experiences, and cultural socialization and norms. How we attribute and interpret clients' responses will be affected by our own ideas and experiences of the meanings of different responses. Observation is useful in case conceptualization to the extent to which we can and do make useful attributions and interpretations about clients' responses and the relation of these responses to the client's problems and change process. Obviously, it is considerably less useful when our attributions or interpretations are not tailored to the particular client and his or her contexts.

EXERCISE 6.8 ■ It is helpful to become aware of our own ways of nonverbally expressing our feelings, as well as ways that others may vary. In small groups, answer each of the following:

1. When I feel angry, I verbally _____ (say) _____ and non-verbally _____ (express) _____ .

2. When I feel sad, I verbally_____and non-verbally_____.
3. When I feel scared, I verbally_____and non-verbally_____.
4. When I feel hurt, I verbally_____and non-verbally_____.
5. When I feel happy, I verbally_____and non-verbally_____.
6. When I am anxious, I verbally_____and non-verbally_____.

After each member of the group has completed this exercise, discuss the differences in responses. How might you and your peers differently interpret particular observed behaviors? As we get to know people, it is usually easier to understand the meaning of our observations and to place these in the context of the clients' contexts, rather than our own contexts.

EXERCISE 6.9 ■ Choose someone in the class whom you do not know very well. Observe this person discretely for the first half of the class. What do you think they are feeling? Thinking? How are they relating with others? At the end of class, check out your observations with this person to see how accurate your observations were. An alternative exercise is to go sit in a cafeteria or park with a partner from class. Separately, each of you can observe several different people. Describe what you think those people may be thinking, feeling, or communicating to others. Compare your answers with your partner's answers.

Information from Counselor–Client Interactions. How we respond to clients and our observations of their responses to us are another important source of information for conceptualization. The responses that the client elicits from you as the therapist are important information about how the client interacts with others; the responses he or she provokes in others more generally; and his or her interpersonal style, strengths, and challenges. How the client responds to us and how we respond to them also tell us about how the client perceives us, even if the client does not directly tell us about this. If we look at the helper-client relationship as continuously circular, we will be able to have some awareness of what we bring to the relationship at any given moment and to separate this from what the client brings. How aware are we of how our own mood states and current contexts affect our observations of the client? If, for example, we are preoccupied with one of our own personal issues, we may not be as fully present with the client as we would like to be. This awareness is essential to enable us to understand how the client and helper continuously affect each other and use this information to deepen our conceptualization of the client. Do we see any patterns in how we respond to the client and how he or she responds to us? How does this particular client make us feel and what does this mean? How do we think the client is seeing us?

Don was a 19-year-old college student who came to therapy because he had recently broken up with his boyfriend who had stated that Don was "incapable of intimacy." Don wondered if this was true and said that he came to therapy to "figure out how to really be close to someone." As a secondary concern, he felt that he was not as invested in his college studies as he would like to be. After the initial intake sessions, the therapist felt increasingly bored in the sessions and found her attention wandering as Don spent most of the time discussing feeling "really bummed out" about getting Bs rather than As in his classes, detailing his approach to school and specific homework assignments, and discussing ways to improve his approach to studying.

The therapist's boredom was a signal to her. If Don was indeed so distressed about his schoolwork, why wasn't this intense emotion engaging her? When this question came to her, she considered more carefully her observation and experience of Don in sessions. She realized that Don *said* he was "really bummed out," but that this feeling did not seem present in the moment and that he reported on his feelings with a matter-of-fact tone. As part of the case conceptualization, the therapist developed a hypothesis that Don's distress about his schoolwork was not something that fully engaged him emotionally and was therefore not something that engaged the therapist emotionally. She also wondered whether Don was interacting with her the way he had with his boyfriend, avoiding talking about things that were really emotionally important or that would make him emotionally vulnerable. The therapist conceptualized that for Don, the discussion of his schoolwork was "safe" and a way to maintain an interpersonal distance.

Therapists often intuitively modify their approach in response to clients' unique presentations and needs. Becoming aware of the nature of these modifications can help us conceptualize clients more fully. Do some clients need more structure than others? How is trust developing? How do we respond to clients' resistance? For example, do we avoid sensitive topics or do we try to introduce them in a timely way that is helpful? How do we know when and how to confront? To provide support? To utilize silence? How do we assess the client's and our own problem-solving styles? What are the dependency/independency; power/lack of power; internalization/externalization dynamics of the client, of our self, and of the interactions between us?

The challenge in evaluating and using information from interactions is determining whether our responses to the client are actually related to the client's presentation and issues. As demonstrated in the case of Don above and discussed in Chapters 4 and 5, these responses *may* be primarily based in the client's presentation, either in the real relationship or in the client's enactment of the transferential relationship. However, our response to the client may also be related to our own mood, feelings, interpersonal strengths, challenges, and so forth. It may be a response that is based on our own countertransference, in the more narrow sense of the feelings and responses towards the client that are related to our own "unfinished business."

EXERCISE 6.10 ■ Over the next few days, observe your responses to a significant other in your life. Identify your thoughts and feelings when with this person and consider how you respond and what variables contribute to your responses. What are the relational variables, the personal variables of you and your significant other that contribute to your pattern of responding? How might you respond differently to the same interactions if they were with a different person?

If we are struggling with a problem that the client also struggles with, it may be challenging to see beyond our own struggle to the client's actual experience. The discussion of Cara in Chapter 5 explores some possible aspects of this challenge. Another example is the case of Beth, a 16-year-old girl who came to therapy through a court order. She had allegedly assaulted her mother's friend with a knife, although Beth reported that this was an accident while she was cooking, gesturing with a knife while unaware that the friend was next to her. Beth's mother, who had supported the friend's charge against Beth, suffered from episodic bipolar disorder. The mother of Beth's therapist also suffered from episodic bipolar disorder. In therapy, Beth described a time when her boyfriend had come to the home that she shared with her mother and beaten Beth in the yard while her mother simply looked on. While hearing this, the therapist was outraged and angry with the mother. Because of this, rather than exploring Beth's response, the therapist asked Beth directly if she was angry with her mother. Beth responded, "No, she's sick, she can't help it." The therapist, who had frequently excused her own mother's emotional abuse in this way, did not know how to respond and felt helpless and angry. In supervision, the therapist was able to explore how her response of helplessness had a major basis in her own experiences with her mother and not in her actual relation with Beth or in Beth's relationship with her mother. The therapist was also able to explore how her anger was related not only to real feelings of caring for Beth and a healthy evaluation that parents should not allow their children to be beaten in front of them, but also to the therapist's own anger at herself and at her own mother.

How do we know that our feelings towards the client are not more our issue than the client's? There is a need for continual self-examination, and often we need to consult with a supervisor or colleague about persistent feelings of ours in response to a particular client to ensure that we are not projecting our issues onto the client.

EXERCISE 6.11 ■ Think about some unfinished business you might have with a family member or significant other. For the next week, jot down your thoughts and feelings about this person and the issues that may be unresolved. Imagine what you would like to do and say, why you do not do or say anything, and how this may affect current professional and personal relationships. What would it be like to see a client with similar issues?

EXERCISE 6.12 ■ Think about times you have been with a client (or a friend or colleague, if you have not yet started seeing clients) and have been distracted, bored, or over or underreacted. From your perspective now, how can you better understand these times?

Your preferences on the relational dimensions will likely have a strong effect on the extent to which you attend to information from the client's responses to you and your response to the client and the attributions you make about these responses. If you see the relationship as central and as a major creator of positive change, you will prioritize information about reciprocal relational responses in your case conceptualization. Often, our responses are a mix of "real" and "unreal"/transferential relationship aspects, just as clients' responses to us are a mix of these aspects. If you emphasize the transference relationship, you will attend to and make hypotheses about transference responses, rather than seeing these responses as related to the current relationship between you and the client. The extent to which you attend to power differences and structural power analysis will affect how much you consider the interaction of your social status and role power with those of the client in your conceptualization.

Secondary Sources of Information

Secondary sources of information come from other people in the client's life, previous records, the referral source, the media, and so forth. Many therapists feel that it is best to not consult these sources of information before meeting the client so that you may form your own impressions untainted by other information. We tend to agree, although we recognize that in many therapy settings, it is a standard procedure for a client to be seen in triage or intake by someone other than the person who will become the primary therapist. If you have information about a client prior to actually seeing him or her, it is important to try and hold this information very tentatively, so that you may develop your own view of a client without imposing preconceptions. However, any information you can obtain after you have met with the client (and with his or her permission, of course) is important in obtaining a more complete picture.

Collateral Information In some situations, clients come to us with school, medical, legal, or previous psychological records. As mentioned, it is frequently desirable to meet with the client and learn from them directly about their concerns before referring to collateral information. If, however, the treatment context is very short-term, collateral information can offer an important "head start" on basic information gathering. In some instances, it is not only past records that are important, but you may want to solicit input (with the client's consent) from other health care providers or external sources. It may be advisable to recommend neuropsychological, personality, or cognitive testing or a psychopharmacological consultation (see assessment, below) in order to understand fully the client in different contexts. For example, there have been times when clients come in for treatment of anxiety or depression and a physical exam reveals an underlying medical condition. Or we want feedback from the school system about their perceptions of a child's difficulties and behaviors with peers. A couple or family session can also provide helpful differing viewpoints, and firsthand observations of a client in relationship systems provides another facet of the client's interpersonal interactions. Notes from a previous therapist can be helpful or not so helpful; how we perceive and interpret this collateral information depends on our orientation and perspectives.

We often share the information we obtain with the client—in a respectful manner and at an appropriate time—after assessing how this knowledge will affect the client and the therapeutic relationship. It is also important to remember that in many states clients have the right to access their complete records. In our experience, however, clients rarely do this and are more likely to rely on what the therapist chooses to share with them. Specific details or "jargon" may not be helpful to the client, but there are likely to be inconsistencies and contradictions between what the client tells us and what others inform us; we can gently and empathically explore these differences with the client, rather than judge them punitively or assume that the client is distorting. Everyone has a "truth" and there is no one truth. Each bit of information is a piece of the puzzle we are trying to put together.

Steve suggested that his wife join him for a session, explaining that he thought his wife's views would be helpful. The therapist agreed as she typically held at least one meeting with significant others or family members of the client in order to gain a broader perspective and to directly observe significant relationships. When Steve's wife joined him for the third therapy session, the therapist was able to compare her perceptions about Steve's presenting problems with his. It was clear to both Steve and his wife that he was unhappy at work and at home. The therapist observed a supportive, strong, marital bond. Steve's wife wanted to be able to help him with his work stressors but did not know how. Over the next few individual sessions with Steve, it became clearer to the therapist that his symptoms of depression and feelings of stress with his family were the outcome, not the source, of his unhappiness at work. A treatment plan focused on changing Steve's core assumptions about his adult roles and responsibilities.

Yuki came to therapy after a thorough physical exam and testing related to her complaints of headaches, sweating, and racing heart, which was important in order to rule out any underlying or co-existing medical condition. If she had not already had a physical, her therapist would have recommended one. We must always be cautious about assuming that physical symptoms are "psychological"; there are many illnesses, such as fibromyalgia, Lyme's disease, Epstein-Barr, multiple sclerosis, and Parkinson's, where it appears as if the early symptoms are psychological and it may be years before the actual diagnosis is made.

Formal Assessments and Other Consultation There may be times when we believe that more formal assessment, such as testing, will be helpful in understanding the client and his or her issues. There are theoretically based formal assessments that some clinicians use themselves (rather than making a referral for an assessment), such as Lazarus's (1989) Multimodal Assessment which focuses on the client's functioning in BASIC-ID areas (Behavioral, Affective, Sensation, Imagery, Cognition, Interpersonal relationships, Drugs/biology) areas and the Adlerian Life Style Assessment Forms which gather information about special characteristics, attitudes, family, gender, and other sociocultural factors of the individual. There are also symptom based assessments such as the Brief Symptom Inventory (BSI) or the Symptom Checklist-90 (SCL-90) which assess the presence and severity of different kinds of symptoms reflecting diverse diagnoses, or more specialized symptom assessments such as the Hamilton Rating Scale for Anxiety or the Beck Depression Inventory (BDI). Many clinicians find these tools helpful, either as a standard assessment instrument used with all clients, or to further explore specific issues that seem relevant to a given client. In addition, some settings have their own pre- and post-treatment assessment forms.

Other times, we may want to refer a client for a formal assessment. For example, further contact with Steve raised a question of whether he may have been suffering from undiagnosed Attention Deficit Disorder (ADD), which affected his focus at work and his interactions at home. After much discussion about his schooling background (his family had moved a lot and he went to seven different schools from kindergarten to high school) and his difficulties focusing, we decided to have formal testing prior to a psychopharmacological consult with a psychiatrist. Up until this point, Steve's primary care provider had been prescribing an SSRI medication for his depression. Although Steve found the anti-depressant medication helpful, the therapist's usual practice was to have a psychiatrist prescribe and monitor medication rather than the primary care physician. However, prior to seeking this psychopharmacological consult, she thought that the ADD assessment would be useful. The results of the testing did indicate a moderate level of ADD and the consulting psychiatrist prescribed medication for the ADD and changed the anti-depression medication to better address the anxious aspects of Steve's depression.

Several years ago one of us worked with a couple where the conflict was focused on the wife's inability to interact effectively with others; in particular, they had a lot of conflict about the care of their severely disabled child. In observing the wife's interactions with the husband and the therapist, the therapist felt that the wife seemed to be unaware of social cues and expectations that most people would likely be able to see. After three sessions, the therapist hypothesized out loud that the wife might have non-verbal learning disabilities and described that condition. The wife started to cry, saying "Thank you! No one has ever understood." A neuropsychological assessment confirmed this hunch and the focus of the treatment became educating the couple about this disability, teaching social cue skills, and developing ways the couple could work together to care for their child given each person's abilities and limitations.

There are often economic constraints on obtaining formal assessments, and the clinician may have to prepare a substantive proposal to the third-party payor or participate in meetings in order to advocate for obtaining these services. There have been many times when we have attended Core Evaluation meetings for school age clients and have been advised by the school psychologist to request more comprehensive assessment for the client than the school is able to provide.

There are also many arenas of a client's life where we do not have access to secondary sources of information. Typically, this includes information from the work system, family-of-origin, and social networks. However, it is always important to query the client about all aspects and contexts of his or her life, and sometimes we might ask the client to obtain feedback from peers, supervisors, and friends to test out the client's hypotheses about his or her own functioning in these settings.

ORGANIZING AND INTEGRATING INFORMATION (PRE-CONCEPTUALIZATION)

As we gather information from the client, and from our observations of and interactions with the client, we must frequently "decode" the meaning. One part of "decoding" is "reading between the lines" to understand the meaning that may be implicit, rather than explicit. This is related to considering not only what is expressed or observed, but also attending to what is *not* said, done, or felt. As previously noted, there may be differing levels of congruence between the client's verbal, nonverbal, content (literal), and analogic (meta-communicational or process) messages. Part of the therapist's expertise is the skill to consider the different pieces of information in relation to all the other pieces. In decoding the meaning, we are *making connections* between pieces of information that illuminate things the client may not yet be aware of. This is a fundamental counseling skill that enables the therapist to form a more accurate picture of the client and his or her circumstances.

Organizing information from various sources is the first step towards integration in order to move towards an effective conceptualization. One way to approach organization is to ask particular questions that help us pull information together. The relative emphasis we put on these questions and the ways in which we answer them are frequently affected by our dimensional preferences and related theoretical orientation and theory of change. As we have noted before, most therapists focus on clients' presenting problems, so the organizational questions they use to integrate information will reflect this focus:

- What is causing the client most distress?
- What do you see as the client's most pressing problems (this might be different than what is distressing to the client)?
- What contextual issues seem most relevant to understanding the distress (consider different levels and aspects of context and their interactions)?
- How are the client's problems or seemingly problematic ways of thinking, acting, feeling, or relating serving the client?

- What are the client's strengths? What resources are available to the client? Internally, relationally, systemically?
- What barriers exist for the client in relation to making change? Internally, relationally, systemically?

The answers to these questions are frequently included in the "presenting problem" and "background" sections of a case presentation.

EXERCISE 6.13 ■ Think about something you want to actively change in your own life. Review the questions above and apply them to that situation.

EXERCISE 6.14 ■ Consider the questions above in relation to Steve, Marie, or Yuki. How do your answers reflect your own theoretical orientation and dimensional preferences? Considering these broader organizational questions can also help us highlight areas where we might need to gather more information. What do your answers tell you about what additional information you would seek if you were working with one of these clients?

CONNECTING INFORMATION TO THEORY: MOVING TOWARDS CONCEPTUALIZATION

Case conceptualization moves beyond organizing information in ways that are influenced by theoretical orientation to integrating information into an overview for which theoretical orientation is a foundation. Case conceptualization, therefore, asks about (a) how and why the presenting problem developed for this particular person, (b) for whom is it a problem, (c) what maintains it and what are the ingrained patterns associated with it, and (d) what is the best approach to positive change. Case conceptualization understands these answers to be informed by theoretical orientation. It also moves beyond a relative focus on the presenting problem to more fully consider the whole person in context and the prospective experience of that whole person within the therapy. In order to do this, case conceptualization considers more broadly the client's view of self, the world, relationships, life plan, and so forth. In conceptualizing clients, we therefore generate hypotheses not only about the presenting problems, but also about the client's overall functioning, short-term and longer-term goals and objectives, and treatment possibilities, including responsiveness to the therapeutic relationship. This is a complex process that becomes less labored with experience.

Although we need to conceptualize clients wholistically (not just their problems), we also need to make choices about what is central to this understanding and to the process of change that is the goal of therapy. There is inevitably a lot of information that you have about the client that seems related to understanding who they are, but may be less related to helping them change in positive ways. What is most important in understanding the client is what is directly related to the goals of therapy. So you not only need to organize and integrate the information you have, but also prioritize (and frequently cull or condense) this information for your case conceptualization in relation to the possible goals of change. This is so the case conceptualization can guide you towards treatment planning, intervention implementation, and outcome evaluation.

Because case conceptualization is founded upon theoretical orientation, different clinicians with different theoretical orientations may conceptualize the same case differently. Therapists differ not only in theoretical orientations, but also in the personal values, experiences, preferences, and interpersonal styles that affect our orientations and our approaches to being a therapist, as we have discussed in earlier chapters. However, theoretical orientation should be a framework to help us integrate information and plan interventions; we should be wary of this framework becoming a constraint. Some cases lend themselves more to one type of theoretical orientation within a case conceptualization.

This is why we advocate for therapist's awareness of their choices in dimensional preferences and related theoretical orientation. As we formulate our story of the client, based on an integration of observations, elicited data and information, and assessment of our experience of the client, we should also be able to identify how our theoretical formulations inform our thinking. Theories are the maps, the guidelines that shape our affect, cognition, and behaviors. Case conceptualization is a clear presentation of how these maps help us to understand a particular client, his or her problems and contexts, and our relationship with him or her in order to provide helpful treatment. We cannot take the unique person of the clinician, or the unique relationship between a particular client and therapist, out of the conceptualization. Thus, there is no one "right" way to conceptualize a case. It is not the content of the case conceptualization that matters as much as the process of integrating the pieces and exploring the relations between these pieces.

Research supports the importance of case conceptualization in developing effective treatment plans. As a result, there have been some instruments developed to measure the problem-solving and treatment planning of mental health professionals (Sperry, 2005; Lee, 2005; Weber, 2001; Sharpless & Barber, 2009; Petti, 2008). Most studies that use control and intervention groups show that attention to case conceptualization results in higher scores on the various instruments for the intervention groups, regardless of specific orientation. However, there is great heterogeneity in the training and experience level of trainees and professionals; contexts differ; research methods differ in terms of qualitative and quantitative measures; and cognitive styles vary. It is difficult to measure abstract thinking, inferences, and intuition.

Thus, researchers are increasingly thinking about effective ways of teaching case conceptualization knowledge and skills at different levels and experimenting with different supervision and teaching formats. Although each theoretical school promulgates its own ideas and meanings about case conceptualization, it is our contention that we can learn from each school and develop and broaden our conceptualization capacities with continuous supervision (formal, peer, informal), experience, and new learning. As instructors of graduate mental health trainees, we are acutely aware of the challenges associated with fostering these capacities. Increasingly, trainees are working with a much broader range of presenting problems and client diversity than previous generations, and reciprocal learning by trainers, teachers, and trainees is inevitable.

EXERCISE 6.15 ■ What would be the best methods to teach you case conceptualization skills now that we have reviewed what case conceptualization is? What kinds of teaching and supervision are effective for you? How do you think you can improve your learning and training?

SUMMARY

In this chapter, we have focused on how you collect and organize information from which your initial conceptualization will be developed. We have discussed how you observe your interactions with the client; how you hear and make meaning of (assess) his or her presentation, self descriptions, and perceptions of their problems; how you utilize collateral information from previous records or other significant others in the client's life; and how you might organize information in order to begin a process of an integrative case conceptualization.

In Chapter 7, we will return to the case of Nancy to illustrate how information about the client and his or her contexts may be integrated with theoretical orientation to develop an initial case conceptualization. We will also explore treatment planning, goal setting, and diagnosis in relation to conceptualization. Although in this chapter we have focused primarily on initial information gathering and in Chapter 8 we will focus on initial case conceptualization, it is important to remember that case conceptualization is a continuous and cyclical process. The therapist is continuously gathering information, assessing new issues, and reassessing issues that were previously considered in light of new information. The client's responses to treatment are, themselves, information to be added to this continuous process. Thus, we must consider how we *continuously* think about possibilities and go beyond snap judgments or diagnostic ruts. What are the relational meanings that shape the experience of the moment, and how are these integrated into understanding the whole client in context? How do we *reassess* our story about the client's story? How do we continuously assess which of these understandings should be guiding contexts and which should be primary foci in treatment planning and intervention? In Chapter 8, we will take up this cyclical process.

REFERENCES

Lazarus, A. (1989). *The practice of multimodal therapy*. Baltimore, MD: Johns Hopkins University Press.

Lee, D. (2005). *Counselor development in multicultural case conceptualization*. Dissertation Abstracts International 66, 3415. http://search.ebscohost.com

Miller, K., Martell, Z., Pazdinek, L., Carruth, M., & Lopez, F. (2005). The role of interpreters in psychotherapy with refugees: An exploratory study. *American Journal of Orthopsychiatry* 75(1), 27–39.

Neville, H.A. & Heppner, M.J. (1999). Contextualizing rape: Reviewing sequelae and proposing a culturally inclusive ecological model sexual assault recovery. *Applied & Preventive Psychology* 8, 41–62.

Petti, P. (2008). The use of a structured case presentation examination to evaluate clinical competencies of psychology doctoral students. *Training and Education in Professional Psychology* 2(3), 145–150.

Sharpless, B. & Barber, J. (2009). A conceptual and empirical review of the meaning, measurement, development and teaching of intervention competence in clinical psychology. *Clinical Psychology Review* 29(1), 47–56.

Sperry, L. (2005). Case conceptualizations: The missing link between theory and practice. *The Family Journal* 13(1), 71–76.

Tribe, R. (2007). Working with interpreters. *The Psychologist* 20(3) 159–161.

Weber, S. (2001). Evaluating the ability of students to articulate case conceptualizations: A qualitative study of two groups of group supervision. *Dissertation Abstracts International Section A*, 61, 3072. http://search.ebscohost.com

7

Conceptualization, Treatment Planning, and Diagnosis

As we organize the information we have gathered from clients, we integrate it with our theoretical orientation to begin developing a case conceptualization and start the process of formulating more specific treatment objectives and methods of attaining them.

INTEGRATING INFORMATION AND MOVING TOWARDS CONCEPTUALIZATION WITH NANCY

Let's begin by returning to the case of Nancy and presenting a more detailed analysis of how her therapist initially conceptualized her within context.

EXERCISE 7.1 ▪ Return to the initial presentation of Nancy in Chapter 1. This case presentation includes information about Nancy from many of the different sources discussed in Chapter 6. Consider what you know about Nancy from the following sources, based on the case presentation:

- Information from Nancy's direct expressions about the presenting problem
- Information from Nancy's direct expressions about her contexts

- Information from the therapist's observations of Nancy
- Information from Nancy's interactions with the therapist

In addition to the information already presented about Nancy, the therapist gathered information about this client from observations of her own reactions to Nancy over several sessions, and from sessions with Nancy's parents.

Although Nancy was never late and never missed a session, after the first two sessions the therapist was aware of her struggle to engage Nancy. Nancy seemed unwilling to explore her own feelings and actions, usually blaming others in her life. The therapist found herself annoyed with Nancy for what felt like entitlement and lack of consideration of others. At the same time, it was clear that Nancy was unhappy and in pain. To overcome her own feelings of annoyance, the therapist found it helpful to empathize with Nancy's obvious pain and anxieties, particularly her feelings that others did not truly care for her. Nancy responded particularly well to this empathy, becoming more relaxed and engaged.

Another example of the importance of validation to Nancy was in relation to the family therapy session. During this session, Nancy's parents expressed resentment about her complaints, and her requests for material goods was a topic of open conflict. Nancy's parents emphasized that they were "self-made" and had no debt. Nancy's behavior during family sessions was noticeably different, in that she seemed quite angry rather than sad and tired as she usually was. The therapist also observed that Nancy interacted with her parents in a belligerent manner, particularly with her mother. Nancy's mother was similarly openly angry and critical of Nancy. The therapist noted that the parents' reactions to Nancy's verbal and nonverbal behaviors seemed to be based on feelings similar to ones that she (the therapist) often had. The difference was that the negative reactions of Nancy's parents were constant and enacted while the therapist's annoyance was episodic and did not override the relational communication of a basic empathic caring for Nancy. This perception was also related to the therapist's observation of her own struggle to empathize with Nancy's mother. While she understood the mother's concerns and annoyance with Nancy's demands, Nancy's mother seemed to have little sense of her distress or pain, in spite of the fact that this was the reason for the session and the therapy.

When the therapist validated Nancy's feelings about her mother's anger and insensitivity in a later individual session, Nancy began to participate more actively in sessions and no longer complained about transportation. From her observation of Nancy's response, it became evident to the therapist that Nancy looked to her for this validation and, with time, Nancy was able to tolerate the therapist's suggestions that she look at her own part in interpersonal conflicts. Likely, this was the first trusting relationship that Nancy had developed and was able to sustain. She demonstrated her commitment to the therapy by appearing on time, by engaging throughout the sessions, and by opening up her thoughts and feelings more frequently.

EXERCISE 7.2 ■ Now, consider the organizational questions from Chapter 6 in relation to Nancy:

- What is causing the client most distress?
- What do you see as the client's most pressing problems? This might be different than what is distressing to the client.
- What contextual issues seem most relevant to understanding the distress? Consider different levels of context.
- How are the client's problems or seemingly problematic ways of thinking, acting, feeling, or relating serving the client?
- What are the client's strengths? What resources are available to the client? Internally, relationally systemically?
- What challenges exist for the client in relation to making change? Internally, relationally, systemically?

As we discussed earlier, these questions can help organize the information necessary for conceptualization. At the same time, organizing information in relation to the problem or distress can shape the way we think, so we need to be careful about how this organization might lead to less attention to contexts, strengths, or other important experiences of the client. Furthermore, in decoding and assessing client information, we are always asking ourselves what might be the more implicit meanings of the client's verbal messages. We also attempt to assess congruence between verbal and nonverbal behaviors, between our understanding of the client's awareness and intentions and our experience of the client's relational effect on us. Thus, we need to assess our own cognitive and affective reactions to the client's information. This is necessary in order to determine the client's meanings and underlying cognitive schema—how he or she perceives, experiences, and makes meaning of themselves, others, and events.

The information we gather from multiple sources is then integrated with our theoretical understandings of development and change. Below are the therapist's initial thoughts about conceptualizing Nancy. Note that the way the information is presented here reflects more than just "the facts," but also integrates the therapist's theoretical understanding and emphases on the therapist's preferred dimensions we have discussed in Chapters 2 through 4.

Nancy's perception of her distress. Nancy described being most distressed by her "nervousness" which the therapist also observed in her kinetic behavior. She also expressed distress about the intense interpersonal conflicts she experienced, especially how her parents do not support her and "deprive" her of things she deserves that they could "easily" provide.

The therapist's views of Nancy's most pressing problems. Nancy's interpersonal conflicts at home and with her suitemates at college shared a common theme: she felt victimized and deprived because she felt that others were not behaving the way that she expected. Nancy verbalized her core thoughts of entitlement and resisted any challenge of these thoughts. The therapist conceptualized her presenting problems as related to core cognitive schemas of "Things should be the way I want them" and "I should have what I want when I want it." Nancy did not seem to have much self-awareness, and she externalized all her difficulties on to others in all of her contexts. Nancy blamed her suitemates, her family, and her teachers for any discomforts and difficulties, and she was unable to consider that she played any role in these interactions contributing to conflict. When the therapist asked questions about the needs and feelings of others in her life, Nancy insisted that only her views were the "right" ones.

The therapist's conceptualization of contextual influences. Nancy's presenting problems were strongly related to her relational contexts. She engaged in splitting behaviors—for example, demonizing her mother and idealizing her father—and she was vigorously critical of the way her parents were raising her disabled sister. Nancy's behavior seemed to fuel family and interpersonal conflicts. For example, she swore and yelled at whoever displeased her, such as demanding that her parents and sister not watch TV, talk, vacuum, or disturb her in any way when she was home. This behavior led to expressions of anger and exasperation from family members; they all felt that it was better when Nancy was not around. Their responses, in turn, affected Nancy's self-concept and self-esteem, which fueled the conflicts and feelings of desperation and anxiety. The therapist believed that Nancy's emotional and social development and distorted thinking likely arose from an insecure attachment to her parents related to their parenting style, the separation forced by childhood hospitalization, and her parents' preoccupation with their disabled child. The therapist also considered Nancy's role in her family as the only child with traditional academic strengths (no learning disabilities) and how these strengths became the source of whatever self-esteem Nancy had while also setting her against her sibling.

The therapist was concerned about Nancy's relationships with her college peers/suitemates and with her secret boyfriend. Relationships with her peers were strongly conflictual and seemed to offer little to challenge Nancy's feelings of alienation and deprivation *or* her beliefs about entitlement. Nancy's relationship with her boyfriend was of concern because of the shame she expressed about this relationship as well as because of the ways in which she described the relationship as primarily for serving her own needs, with little awareness of the man as a separate person with whom relational intimacy might be possible. While the therapist found herself annoyed with Nancy's entitlement assumptions in these relationships, and concerned about how these relationships seemed focused only on Nancy's needs, she also conceptualized that Nancy's conflict with her peers and her clinging to her boyfriend were related to feelings of alienation and deprivation from her family, particularly her desperate seeking of attachment from a non-affective mother.

Nancy's values and attitudes were shaped by the materialism of her peers and by the larger community values of material, social, and educational achievement. The community that Nancy grew up in was very affluent, materialistic, and privileged, which the therapist was aware of from treating other clients in the same community. Because Nancy's parents had told her that they ranked in the lower socioeconomic quartile of this community, she reported that she was sensitive to "being different" and "not having as much as everyone else." Nancy attended a private college where similar privileged values prevailed. Thus, the affluent community in which she was raised and the college context in which she currently lived also shaped Nancy's values and attitudes about money and privilege. Nancy had always compared herself and her privileges to her peer group, which contributed to her stated feelings of deprivation.

How the client's problems serve the client. Nancy's ways of thinking (that others should be the way she wanted them and meet her needs) fueled her feelings of victimization and blaming others for her difficulties. This meant that she did not have to take responsibility for changing, but could place that responsibility on others. It also meant that she could be angry at others' lack of response and avoid her fear that she was actually undeserving of their care. While Nancy hated being the scapegoat in her family, it did get her attention and she was the "center" as well as "target" of all family interaction. In addition, her behaviors were also protective because even if Nancy had to confront that she actually played a part in how others responded to her, she could attribute the experienced rejection to the behavior, rather than to a confirmation of her fear that she was, at the core, unlovable.

Strengths and resources. Nancy had considerable strengths in her determination to succeed in academic and career-related efforts, including her desire to "make things better." Even though she expressed some ambivalence about therapy, her behavior indicated that she was strongly committed. An additional strength was Nancy's ability to create a working alliance. She was able to respond to the therapist and be affected by her. For example, Nancy could experience and respond to the therapist's validation of her feelings such that the tenor of the relationship between Nancy and the therapist changed. Furthermore, in spite of the conflicts, Nancy did have social skills that enabled her to have social interactions with peers. She was not socially isolated or withdrawn, although it was not clear to the therapist how much these relationships were emotional resources for Nancy. These relational strengths suggested that Nancy had some capacity for attachment and seemed to be actively craving a genuine connection. Nancy also had financial and practical resources that enabled therapy and other opportunities for positive experiences in her life, even if she did not experience herself as having these resources.

Challenges. It is possible to see the problems themselves or the developmental foundation of the problems (if your theoretical orientation conceptualizes this) as challenges to change. In some ways this is true, particularly if the problems have become a pattern that serves particular functions for the client or for the family system. But there are frequently other types of challenges as well. For

Nancy, one challenge was her limited ability to take perspective, to step outside of her experience in order to consider the intentions or effects of that experience. Nancy struggled with taking perspective in relation to her own experiences as well as those of others. Nancy's attachment history is an additional challenge, related not only to the conceptualization of the presenting problem, but also to the therapist's understanding of how best to create change. The therapist was aware that this might be a challenge to change because of the way in which Nancy's struggles with attachment might affect the therapeutic relationship itself, not only her relationships with others. This is particularly important because this therapist usually placed a strong emphasis on the relationship. Finally, within the family system, the therapist wondered if Nancy's role as "demanding" and a trouble maker might actually serve the family in some way, and therefore, if Nancy might encounter some pressure from her family *not* to change.

In the above discussion of Nancy's case, you can see how the therapist's initial conceptualization was influenced by psychodynamic theory (attachment theory, relational theory), cognitive–behavioral theory, and family systems theory. Her approach also included elements of constructivist theory in that she focused on Nancy's story, her mother's story, and eventually her father's story, while encouraging Nancy to think about how she might want to change *her* story. The therapist generally has dimensional preferences for a less directive, more active approach, with varying preferences for structure depending on the session. She is more individually and family focused, but pays strong attention to the cultural and social systemic context (for example, in considering the influence of the affluent and privileged community). When treating individuals, her focus of change is more strongly oriented initially towards affect and cognition, although change in behavior is seen as a major indicator of ongoing change. This therapist sees the relationship as central, both as a means and an end, and attends to both the real and the unreal aspects of relationships. She pays strong attention to process, but does not actively share her process observations as means to change, at least not initially with Nancy.

EXERCISE 7.3 ▪ Imagine that a therapist influenced by a different theoretical orientation, with different dimensional preferences, was treating Nancy. How might that therapist's understanding of Nancy be different than what is detailed above?

INTERACTIONS OF INFORMATION GATHERING WITH EARLY CONCEPTUALIZATION AND TREATMENT PLANNING

Because intervention begins immediately, treatment planning also begins even before we take the time to consider longer-term goals for the client or to formalize planning in a write-up. Even as we use the first few sessions to gather information and develop our conceptualization of a client in context, we are already engaging in intervention and making decisions about how we will approach the client in these sessions based on our developing conceptualization.

Initial Treatment Planning during Information Gathering with Nancy

At this point, let's share with you how the therapist thought about treatment for Nancy during these first few sessions in order to illustrate how a therapist might choose initial goals *while* information gathering and building an initial conceptual understanding of the client. Our point here is that we are never just gathering information; we are also beginning the process of conceptualization and change. Because of this, we need to be mindful about our choices and goals even in the very first few sessions.

Nancy's therapist's initial priority was developing trust through fostering an empathic, supportive relationship. Although a working alliance is always an important first step, the initial relational focus of the therapist aimed for more than a basic working alliance (which, if you recall, may have less of a feeling of being validated or fully accepted). It was apparent that Nancy did not have a trusting, secure relationship with anyone. The therapist was also aware that she was an older woman, similar in age to Nancy's mother; Nancy's issues with feelings of invalidation from her mother made a supportive therapeutic relationship particularly important. In addition, the information from Nancy indicated that she frequently felt a lack of control and fulfillment. From a theoretical perspective, the therapist adopted a client-centered relational model as her primary strategy for collecting information upon which to develop a case conceptualization and treatment plan, reflecting her simultaneous goal of relationship building.

The validation of Nancy's own experience also seemed particularly important. Although Nancy seemed to be sure that her experience was "right," the therapist questioned whether Nancy genuinely experienced her needs and feelings as valid, given her extreme protest accompanied by her reluctance to consider the perspectives of others. Perhaps Nancy was simultaneously protecting against anticipated invalidation of her own experiences and re-enacting what she perceived as a normative response to others that she had experienced (lack of empathy). Thus, the therapist paid particular attention to Nancy's perception of her major problems, such as her feelings of being a scapegoat. As previously mentioned, the therapist could validate these feelings explicitly after the family session with Nancy's parents. This approach to intervention was guided by existential–humanistic frameworks in order to address attachment issues conceptualized within a psychodynamic framework.

Finally, the therapist's initial treatment conceptualization of Nancy recognized the need to address the cognitive schemas related to Nancy's feelings of deprivation. She began early to utilize cognitive–behavioral approaches to gently challenge Nancy's assumptions about herself, others, and the world. For example, she asked questions such as "How do you think your suitemate feels when you tell her to turn off the radio?" or "Do you think your classmates insist that their mothers do their errands?" In these questions, the therapist was beginning the process of exploring Nancy's unexamined assumptions that her needs should be the central priority of others and introducing Nancy to the possibility of other perspectives.

These were the therapist's initial thoughts about treating Nancy and were implemented during the first few sessions even as additional information for case conceptualization was being gathered. These goals are clearly related to the therapist's emerging understanding of Nancy within her theoretical orientation and dimensional preferences as described above. Treatment planning beyond the first few sessions is a similar process as described here, albeit with usually broader goals.

CONCEPTUALIZATION AND TREATMENT PLANNING

Conceptualization is an interactive and iterative process of understanding the client and relating that understanding to treatment. Thus far, we have focused primarily on how we *understand* the client in context and how that understanding is affected by who we are (and what we believe) as therapists. We have also considered how gathering information interacts with beginning goal setting and intervention. We turn now to consider more fully the treatment plan as an aspect of conceptualization.

Therapists' Intervention Preferences

The strongest influence on your choice and prioritization of goals and interventions for treatment planning should, of course, be the client's experience and needs. But it should be obvious by now that your conceptualization of the client's experience and needs—and therefore the goals related to those—are influenced by your theoretical orientation and dimensional preferences. Your choice of interventions will also be influenced by your theoretical orientation, dimensional preferences, and comfort with different approaches.

Thus, there is a continuous challenge of figuring out whether your choice of goals and interventions is the best match for the client or simply the best match for you. This is why we place so much emphasis on developing awareness of yourself as a therapist; your own preferences on the dimensions we have discussed; your own affinity for different theoretical orientations; and your comfort at this moment in your development as a therapist with different ways of understanding and intervening. Awareness of your preferences should not solidify them in stone or be used to dictate or justify choices, but rather to a) become aware of how they work for you; b) guard against those preferences becoming a default applied without careful consideration; and c) encourage the active consideration of non-preferred understandings and approaches in case these are a better fit for a particular client. Awareness of your knowledge and comfort can also help you identify goals for your own growth as a therapist.

EXERCISE 7.4 ▪ In Chapter 4, we asked you to consider how comfortable or uncomfortable you might be in understanding or conceptualizing in different ways. Here, we ask you to think about your comfort or discomfort in *actually applying* those understandings or bringing them into the therapy session. Think about your experiences as a therapist (or imagine yourself as a therapist, if you have not yet begun). Rank the statements below from 0 (very uncomfortable) to 4 (very comfortable). Some of the questions have multiple parts, so consider carefully how you may be comfortable or uncomfortable with different aspects within the question.
In therapy sessions with clients, I am (very uncomfortable, uncomfortable, comfortable, very comfortable) with…

1. Exploring with the client his or her present experiences and relationships and how these affect the client's functioning (positive and negative).
2. Exploring with the client his or her past experiences and relationships, connections between past and present experiences, and how these affect the client's functioning (positive and negative).
3. Exploring with the client his or her experiences, beliefs, behaviors, or aspects of worldview that are different than my own (cultural experiences, family experiences).
4. Exploring with the client his or her experiences with discrimination or oppression, even if it is from people like me.
5. Being active in a session, that is, speaking frequently with questions or sharing of my thoughts and understandings.
6. Being less active in a session, that is, being silent, waiting to see the direction a client takes without my questions, or keeping my thoughts or understandings to myself.
7. Being directive in a session, that is, directing the client to discuss or focus on specific content.
8. Being non-directive in a session, that is, allowing the client to discuss whatever content he or she wants to.
9. Being didactic, that is, teaching the client.
10. Utilizing structured strategies (two-chair technique, metaphors, imagery, journaling, genograms) to help the client explore content in new ways.
11. Being unstructured and allowing the client to explore experiences however she or he desires.
12. Working to explore and change the ways that clients think.
13. Working to explore and change the ways that clients feel.
14. Working to explore and change the ways that clients act or behave.
15. Focusing on specific changes or goals that are explicitly described and evaluated (results oriented).
16. Discussing the client's relationships and how he or she communicates relational messages in these relationships (process oriented).

17. Discussing the client's relationship with me (the therapist) and how he or she communicates relational messages in this relationship (process oriented).
18. Discussing with the client the "unreal" relationship between us: the client's feelings of transference and how it is affecting the client or the therapy.
19. Discussing the ways that the client's family may affect the client's experiences including the development, maintenance, or change of presenting problems.
20. Discussing the ways that the client has made personal choices that affect the client's experiences including the development, maintenance, or change of presenting problems.
21. Discussing the ways that the client's community or social networks may affect the client's experiences including the development, maintenance, or change of presenting problems and of strengths and resources.
22. Discussing the ways that the client's social context and statuses (race, social class, sexual orientation) may affect the client's experiences including the development, maintenance, or change of presenting problems.

EXERCISE 7.5 ▪ Consider your answers to Exercise 7.4. How would you like your answers to be different three years from now? What are your goals for yourself in terms of professional change and development? How do you think you can achieve these goals? Again, discuss with a partner or in small groups.

You may prefer or be most comfortable with certain types of intervention, but they may not always be in the best interests of the client. We need to be continuously attuned to the client's reactions and responses to our interventions so that we can time and modify them appropriately. On the other hand, there are certain interventions that we may not like or enjoy using, but that may be appropriate for a client's problems. At one time, one of us, who had received some training from Wolpe (1969) in systematic desensitization, found herself treating mostly patients with phobic disorders who responded favorably to this intervention. But she found it a boring intervention and eventually found herself hoping the client(s) would cancel or not show. She decided to see a more varied client population so that she would be able to retain a positive attitude towards using this technique with just one or two clients per week. This meant that she referred many future clients with phobic disorders to other therapists who were also skilled in systematic desensitization. She recognized that the more dynamic and relational interventions she was interested in developing were not necessarily what was needed by these clients at that time.

Treatment Planning: A Journey, Not a Destination

One of the biggest challenges for many therapists (particularly novice therapists) in treatment planning is choosing and prioritizing goals and the interventions you will use to try and achieve these goals with the client. This can be particularly challenging for novice therapists because they may believe that their treatment plan "should" be fully comprehensive and "should" map out a path to complete psychological health. These beliefs are frequently related to an idea of a treatment plan as a prescription for the entire therapy, as a thing that is unchanging and continually applicable. They may also be related to a feeling of responsibility for the client, that because a client is now *our* client, we are responsible for ensuring that he or she is happy and healthy. This feeling of responsibility can contribute to a belief that helping the client is an all-or-none proposition.

Therapy is not like a journey where the environment is fixed and unchanging, the pace is steady, and the destination is known. It is more like a journey where the terrain is constantly changing. The terrain changes as you (and the client) come to know and understand more and as the client's life and contexts change over time. Furthermore, the steps taken along the journey (the smaller changes that therapy facilitates, catalyzes, or creates) change the view of the terrain as well. The changes in terrain may lead you and the client to take a very different path than what was originally foreseen, or even to set a different destination! Because future steps cannot be fully anticipated from the start, there is a need for constant revision of the "map" that is conceptualization and treatment planning.

Thus, it is helpful to understand that a treatment plan (and the larger conceptualization) is always related to a moment in time. It is dependent on the information that we have about a client (and our related conceptualization) at a particular moment, with an acknowledgement that we will inevitably gather more information that will expand or even challenge our current understanding. It is also created with an understanding of the client in relation to time and change. So the question we ask in treatment planning is not necessarily "What would be most helpful for this client?" but rather "What would be most helpful for this client *at this time?*"

A good treatment plan is always tentative and open to modification. Therapy is not like lecture-based classroom teaching or construction plans. We do not develop a step-by-step plan and expect to utilize it unchanging within a session or across several sessions. The journey frequently has detours, side journeys that are catalyzed by unexpected occurrences or changes in the client or the client's contexts. Even an approach that may be seen as "prescribed," such as a manualized therapy, needs to be responsive to new information and changes in the client and the client's contexts. In our role as therapists, we need to be open to the detours that the client's needs demand, to be flexible and adaptive so that we are attuned to the issues and emotional state the client brings to each session and not hold so tightly to our original treatment plan or approach that we ignore what is most pressing for the client.

Furthermore, one does not make such a complex miles-long journey in one step. Although we may write out a treatment plan with broader goals and intervention plans, there are almost always mini-treatment plans for a particular session or time frame in the therapy, even if these are not formally written up or recorded. But even these may need to be scrapped or modified. One of us was appalled when a doctoral student who had entered a training psychoanalysis described her distress the previous day when she went to her analyst and told her about finding her landlady murdered the previous morning and her analyst ignored it and continued to probe the client about her childhood. I urged this student to run, not walk, back to her previous therapist!

Although our treatment plan needs to be flexible, it is an invaluable tool for connecting our conceptualization of the client to our choice of interventions. Developing a treatment plan is an opportunity for us to step back and consider the client *and the therapy* in context. Furthermore, although clients bring up new dilemmas that need attention and direct the therapist from previous planning, this frequently does not mean that your treatment plan is completely wrong. You and the client will likely be dealing with the same major themes, but in different circumstances and contexts that are more closely related to the particular emerging issues.

Developing Goals

The choice of goals for a treatment plan at a given moment in time are usually guided by 1) the client's own stated goals; 2) the conceptualization of the client in context, with an emphasis on addressing the current problems and experiences, as we have described above; and 3) constraints on the therapy from the therapy context, the therapist, or the client.

The client's goals Nancy's stated goals are: 1) to be happy; 2) to achieve academically; 3) to persuade her parents to give her more money; 4) to be less anxious.

EXERCISE 7.6 ▪ What do you think about Nancy's goals? When Nancy says she wants to be happy, what do you think that means and what kinds of questions could you ask to find out? Ask others in your class what "to be happy" means to them in order to see the variation in what that might mean. What goals might you suggest that Nancy consider in addition to or in place of her stated goals? How might you respond to her goals—what questions might you ask?

Clients' goals are frequently quite broad in scope (such as "to be happy," "to be less anxious"). In contrast, the goals indicated by the therapist's conceptualization are usually more specific, addressing what needs to change or how something needs to change in order to achieve the client's goals. It may also be the case that the therapist may not agree with the client's goals, particularly if the therapist conceptualizes the client in a way that suggests that one of the client's goals may actually be related to or supporting a problematic way of thinking, feeling, or behaving (such as Nancy's goal to "persuade her parents to give her more money"). However, the therapist's goals need to be understood by the client as related to the client's goals, or the client will be unlikely to be motivated to continue in treatment. Part of the challenge is considering how to frame goals that respect the client's perspective and communicate them in ways that feel validating (or at least not alienating) to the client. Utilizing the motivational interviewing model presented in Chapter 4 can facilitate collaborating with the client to share in formulating goals and working to achieve them.

The therapist's goals A major point we want to reiterate is that there is no one "right theory" to explain a client's problems; no one "right choice" of goals; no one "right way" to treat a client; and no one "right therapist" for a particular client. As we have discussed, there are multiple intersecting points of view that inform each therapist's thinking about self as helper, client in contexts, and relationships that promote change. Although there is no one right way, treatment planning and implementation should be strongly tied to the understanding that one has of the client in context, informed by the information one has gathered about the particular client, as well as the evidence one has about similar clients and presenting problems. Thus, the therapist's goals are influenced not only by his or her conceptualization of this particular client, but also by his or her knowledge about similar clients, similar types of problems, different types of therapy, and the interaction of these variables.

The therapist also wanted Nancy to be happier, less anxious, and successful in her endeavors. The therapist's conceptualization suggested that the problems related to these goals (unhappiness and anxiety) were related to insecure attachments, which was related to Nancy's belief that others should fulfill her needs and desires, and that she was not responsible for her own happiness, the effect she had on others, or her response to others. While Nancy believed her problems were entirely externally created, the therapist hypothesized that Nancy was experiencing anxiety and conflicts in all of her relationships due to insecure attachments. She further hypothesized that these attachment issues and related cognitive schemas contributed to the difficulty for Nancy in tolerating or regulating her own feelings and moods. Thus, the therapist had goals of helping Nancy learn to regulate her own emotions (including anxiety), to develop awareness of her emotions, and to change her relationship patterns.

EXERCISE 7.7 ▪ Now, consider what *your own* treatment plan for Nancy might be at this point in time. What reasonable goals would you have—in the short-term, mid-term, and long-term? How would these goals be determined? What would your priorities be? Write down the issues you would want to address, the order of priority, and the preliminary interventions you think would be effective to achieve these goals. As you work on this exercise, consider what further information you might want from Nancy. And keep in mind that there are realistic factors that influence the formulation of goals, such as your level of professional development, constraints of resources and number of sessions available, and the client's capacity and motivation for change.

When considering goals for a given client, the therapist's conceptualization of the client's contexts at the multiple levels of individual, family, and structure and the therapist's awareness of his or her own preferences and biases are very important. For example, a therapist who was not attending to the cultural context may unthinkingly set a goal with Yuki (from Chapter 6) to talk to her family and friends more openly about being raped. But this may actually be harmful to Yuki, given her beliefs about her context, and needs considerable further exploration. One of us consulted on a case of an Asian American college student struggling with anxiety and depression related to choosing a major and career path. Her family wanted her to go into a business-related field, but she was interested in humanities. The therapist set a goal of greater autonomy in her decision making, decreasing reliance on her family, and feeling less obligated to please her parents. These goals, however, are not culturally congruent for more enculturated Asian American families and clients. The therapist conceptualized the issue as a conflict between the client and her family when, in fact, there was a conflict within the client who wanted to pursue her love of the humanities and simultaneously wanted to be a good daughter and maintain her connection to her family and culture. The therapist's choice of goals distressed the client, who felt invalidated and misunderstood.

EXERCISE 7.8 ▪ Consider one of the cases we have previously presented: Thanh from Chapter 3; Maria from Chapter 4; or Yuki, Steve, or Marie from Chapter 6. Choose the client that you felt most different from in terms of your familial and cultural background or your sociostructural statuses. Consider what kinds of goals you might initially consider from your own context, if confronted with the issues that these clients face. How might these goals fit or not fit in the specific contexts of these clients?

Constraints on goals The choice of goals and interventions is also related to constraints on the therapy from the treatment context, the therapist, or the client. Constraints from the treatment context might include access to and availability of therapy at a given clinic or organization, types and quality of services allowed or available, costs of treatment if not covered by insurance or other funding sources and financial policies, scheduling flexibility, and so forth. An example of when the treatment context sets or affects the goals might be when a child is discharged from a psychiatric facility due to insurance restrictions before the staff believes he is sufficiently stabilized. Another example may be when a client is not able to receive the type of therapy her problems require because a particular form of treatment is not available at the site to which the client has access. A third example is when the therapist and client need to choose which specific issues will be addressed and which will not because they have only limited time together.

Constraints from the therapist may include issues about the therapist's time and availability. Obviously, there needs to be mutually convenient times for regular appointments. Therapists who are trainees often have constraints due to their limited availability, such as doing clinical work on only specific days or being available only for a few months if the client comes in towards the end of the training year. Finally, the therapists' awareness or lack of awareness of their own inherent biases or limitations can be an issue. These biases could result in the therapist avoiding certain treatment issues, goals, and interventions. For example, if a therapist is uncomfortable talking about sexuality or sexual orientation, she or he might not think to explore these aspects of the client's life. Or if a client's substance abuse issues emerge midway through a treatment for anxiety and marital therapy, a therapist who is unaware or insecure about his or her limitations in treating substance abuse may avoid developing goals about these issues.

Constraints on goals may also come from the client or the client's contexts. Obviously, the client's motivations for seeking therapy and capacity for engaging in therapy despite barriers are critical factors. Particularly if the client is an involuntary one, the therapist needs to be skilled in working through resistances and having the patience to wait for an opening. Other constraints may be related to practical issues such as scheduling difficulties given work and family responsibilities. Transportation and physical access may also be issues; these are also responsibilities of the treatment context, so an interaction of effects exists. If the therapist works in a building that does not have handicapped access, a client may be unable to meet even if the client wants to work with this particular therapist.

Constraints on developing goals for therapy frequently come from multiple sources, or are negotiated through influences from the treatment context, the therapist, and the client. While it can be frustrating to deal with such constraints, it is also an example of modeling compromise and realistic expectations for clients more generally. One of us saw a client during building construction. After a few sessions of having to move from office to office (sometimes at the last minute), the client commented that she was impressed with how unflustered the therapist seemed, and how she always found a place that would work for their meetings. The therapist actually had felt quite flustered and had been fearful that the constant moving was making it seem to the client that she was unprepared or did not care about the client. The client's response was a good reminder that the ability to deal with constraints is frequently a goal of therapy!

An example of interactions of constraints and goal setting is Marie, from Chapter 6, who was referred through workers' compensation for pain and anxiety related to a repetitive movement injury. Some of the goals of that therapy (pain reduction and improved functioning in preparation for return to work) were dictated by the therapy context. Other goals, such as the focus on self-care, were related more strongly to the therapist's conceptualization of Marie. Although the treatment context frequently affects the goals, the influence of the context should not be unquestioned by the therapist, who needs to differentiate what might be best for the client from what is possible for the client in this context. What role should the reimbursement policy play in treatment decisions? What role should the needs of other social or institutional systems, such as healthcare, school, work, or the judicial system, play in treatment decisions? How much latitude do you have in developing a treatment plan in your own treatment setting?

As mentioned, a major constraint related to many therapy contexts is the time available for the therapy, particularly in this day of managed care. A treatment plan *inherently* references a time span for goals and interventions. What we mean by this is that the goals set forth in a treatment plan are related to the therapist's understanding of what time frame is available for the therapy. The goals you set for the treatment plan for a client who only has 10 sessions (due to insurance compensation limitations, organizational policy, relocation of the client or therapist, and so forth) will be different than the goals you set if you anticipate working with a client for a year or two.

The time available for therapy is sometimes related to organization policy, such as a limit on the number of sessions, but most frequently related to financial constraints. Thus, it behooves us in our initial session to explain our agency policies and to discuss financial matters such as inquiring about third party coverage and clarifying when and how co-pays will be collected. If there are a pre-determined number of sessions authorized and there is a possibility that more sessions might be needed, it is important to clarify fees and policies at the onset of treatment so that the goals can be developed with a realistic understanding of how much time will be available. Talking about fees and financial policies can be challenging as many people (including therapists) find talking about money very uncomfortable—sometimes even more uncomfortable than talking about sex! However, it is important that clients are not taken by surprise by issues related to ending or modifying therapy due to financial or time considerations. (See Chapters 5 and 9 for more on this issue.)

EXERCISE 7.9 ■ It can be helpful for you to practice talking about financial issues so that you will be able to help clients with their discomforts and issues regarding money. Choose a partner and interview each other about your financial circumstances now and those of your families when you were growing up. How was money managed and talked about in your families? How comfortable are you today talking about money, and what issues do you think you have managing your money? What does money mean to you? Who and what have shaped your thoughts, feelings, and behaviors pertaining to money? In many cultures, money represents power and can be used to control others' behaviors. How do you relate to this notion?

In addition to negotiating financial constraints, one of the biggest challenges in developing goals is choosing and prioritizing among possible goals. This is particularly challenging when you believe that the client's most distressing problem(s) will require more time to address than you have scheduled for treatment. In this case, it is important to focus on therapy as a journey of many steps and consider which small steps can be accomplished in the time available. When a client presents with so many overwhelming problems in so many domains that it is difficult for a therapist to know where and how to start, the first step is for the therapist to help the client select one manageable goal to work on. Within a supportive relationship that offers some hope and possibilities, the therapist could break this goal down into the smallest elements and help the client to take one step at a time.

It is also important to remember the changing terrain and the effect on the client: change does not happen only in the therapy room. With a limited time frame, it is most helpful to choose a focus that clients can continue to build on after they have terminated therapy. However, even if time may be unconstrained by the treatment context and the number of sessions is not pre-determined, time is still relevant, in relation to considering the kinds of resources (financial, emotional) that the client has.

Positive contextual factors Whereas we have discussed the contextual constraints of managed care and agency policies, a principal benefit may be the focus on therapist accountability. Within these constraints, therapists are required to elucidate short-term, intermediate, and long-term goals for the treatment of whatever the diagnosis is along with specific interventions used to achieve each goal and also to specify criteria for evaluating treatment outcomes. While a lot of record keeping and paperwork is required, clinicians are forced to be specific and clear about what they are attempting to achieve and to provide documentation for what they do and eventually achieve.

Working within an organization, as opposed to having an isolated private practice, may foster therapists' problem solving via team meetings, case conferences, and peer supervision. Organizations may provide opportunities for professional development so that clinicians can expand their knowledge and skill bases.

Obviously, client circumstances can be beneficial as well as difficult. A client whose treatment is supported by significant others and who does not have difficulty adhering to appointment schedules may have more psychic energy for engaging in therapy.

Developing goals with Nancy A major issue related to the treatment context with Nancy was that after the first few sessions, she made a decision to enroll in a 10-week study abroad program in London, for which she would leave in two months. This changed the context of treatment from more open ended to a very short time, at least for the first phase of treatment. When asked why she had decided to enroll in this program, Nancy said that she did not really know, but that she thought it would look good on her resume. She had several worries about going abroad: feeling anxious, having enough money for the program, and whether she would be able to academically succeed. Nancy was also worried about whether she would like the other students; she knew only one other student from her class attending this program and she did not care for this other student. Finally, Nancy was worried that her boyfriend would find someone else while she was away.

All of these issues were related to the therapist's conceptualization of Nancy, but it was clear that the therapist's initial broader goals of emotion regulation and changing relationship patterns were not going to be fully or even majorly addressed within six to eight sessions. This led the therapist to develop an initial treatment plan with consideration of what could be accomplished in this limited time frame as a first phase of therapy and to consider how this new context of international travel could be an experience that was helpful to Nancy's psychological growth. These decisions were affected by Nancy's stated intention to continue in therapy after she returned from England.

The therapist and Nancy together developed short-term goals of preparing Nancy for her study abroad and developing strategies that would enable her to use this time to become less dependent on her mother and her boyfriend. More specific goals encompassed behavioral plans for ensuring that Nancy would learn to do more of her personal errands, manage her own finances, and complete her coursework on time in order to not jeopardize her acceptance into the study-abroad program. Other goals included developing strategies to help Nancy manage her anxiety levels, which she anticipated would increase from the stress of travel and new environments and experiences.

Given the new contextual information of travel abroad and the time constraints imposed on the therapy by its timing, the therapist chose to not prioritize goals related to increasing emotional awareness and tolerance, which she had felt would be central to emotion regulation and changing relationship patterns. The therapist felt that Nancy was too overwhelmed by the challenges presented to her by going abroad. There was also not sufficient time to ensure that enough progress would be made so Nancy would not be left with intense emotions with few strategies or resources to cope with newly emerging feelings. The therapist also chose to set goals that took smaller steps towards addressing Nancy's relational issues, framing the time abroad in relation to independence, but not centering the therapeutic work on the cognitive schemas or relational attachment issues she conceptualized as central to her relationship problems.

EXERCISE 7.10 ■ Review the treatment plan you proposed for Nancy in Exercise 7.7. How would you modify your treatment plan after finding out that Nancy was leaving in eight weeks for her study-abroad program? What would you choose to focus on in those eight weeks? What would you choose *not* to prioritize as immediate goals? How are your choices similar to or different from those of the therapist as described here? What are the similarities and differences in your theoretical orientation, dimension preferences, experience, or conceptualization of Nancy that relate to how your choices and those of the therapist differ?

In summary, choosing treatment goals is a process of considering what is best for this client at this time given the context of therapy. Thus, you could consider:

- What are the client's goals?
- Given your conceptualization of the client in contexts, what do you consider the most important goals for this therapy? Consider your short-term goals (5–10 sessions, for example), as well as longer-term goals. How do these relate? Consider how your goals relate to your conceptualization of the client, your theoretical orientation, and your dimensional preferences.

- What are the practical constraints imposed by the therapy context (time, focus, and so forth)?
- Finally, integrate these thoughts into a realistic treatment plan for this particular therapy, with this particular client, at this particular time.

CHOOSING STRATEGIES FOR TREATMENT GOALS

Once we have formulated a treatment plan stating problems, goals, and desired outcomes, we can consider what strategies are likely to be effective. The goal of this book is not to recommend or describe particular strategies, so in this section we are discussing more general approaches to choosing strategies and different kinds of influences on those choices. You may want to consider the cognitive, affective, and behavioral dimensions separately, or you may want to integrate them. As we get to know clients, we are better able to tailor our interventions to their strengths and style. Similarly, as we get to know more about clients and therapy in general, we are better able to tailor our interventions as we develop new understandings and skills, become more comfortable with different approaches, and increase our awareness of how our theoretical models relate to being in the moment with real people. Thus, our choice of interventions in a given moment is related to the client (of course!) and also both to the developmental stage of the therapy and the developmental stage we are in as therapists.

As discussed in Chapter 1, there is a growing body of literature exploring the effectiveness of different strategies and approaches for the treatment of particular presenting problems or interventions with particular populations. In addition to specific research studies of empirically supported treatments, there is literature that overviews strategies and approaches (for example, Barlow 2010). There is also scholarship that describes different strategies and the varying ways they may be used, adapted, or integrated (for examples, Ballou, 1995; Lazarus, 1997, 2008). This type of scholarship is useful in considering various kinds of interventions and developing "tools for our toolbox."

Also as we discussed in Chapter 1, we believe that conceptualization is an integral part of evidence based practice. Research is a vitally important part of choosing strategies, but should be carefully utilized, as a given client will be both like and *unlike* the samples that were used to establish these guidelines in multiple and intersecting ways. People are complicated and research rarely explores those complications. For example, one of us treated a client with severe obsessive-compulsive disorder (OCD), which has been shown by research to be most responsive to cognitive–behavioral interventions (Franklin & Foa, 2007; Watson, Anderson, & Rees, 2010). However, this client also had a history of severe physical, emotional, and sexual abuse. This history contributed to great difficulty in trusting others (including therapists) and a deep need for validation and support in order to counter a tendency toward intense self-blame and hopelessness. Although the therapist was trained in cognitive-behavioral interventions appropriate to treating the OCD, the client was not at all open to the kinds of structured cognitive-behavioral interventions that have been empirically validated with OCD. We worked together for two years with more psychodynamic and existential–humanistic strategies to address issues of attachment, trust, and self-concept. At the end of those two years, the therapist moved away and the client accepted a referral for a cognitive-behavioral therapy.

EXERCISE 7.11 ▪ Select two of the problems or goals you described for Nancy in Exercise 7.7. What does the professional research literature report about treatments for these problems? In other words, what might be the "best practice interventions" or "evidenced based practice" for clients of Nancy's race, culture, age, and gender with this problem? What *doesn't* the research tell you? How might Nancy be different than the people used in the research you have reviewed?

In addition to research that provides guidelines for choosing interventions for particular problems or populations, there is also literature that explores the complex intersections of therapist, client in contexts, and conceptualization through case study approaches (for example, Gallardo, M., & McNeill, 2009). Your own experience (and that of your supervisors or peer consultants) is also a source of knowledge. However, it is important to remember that, similar to the information you gather from clients, the types of evidence you seek out and attend to—whether from research, theory, case literature, or direct experience—will be affected by your theoretical framework and your dimensional preferences. Thus, therapists' philosophical assumptions lead them to formulate and enact treatment plans and processes that reflect their theoretical frameworks and their personalities and backgrounds.

Each traditional theoretical orientation has a number of techniques, strategies, and approaches to intervention associated with it. Therapists with an integrative orientation often use strategies and techniques from multiple orientations, and part of the choice is related to the particular conceptualization for this particular client. With Nancy, the therapist chose to use progressive relaxation and meditation exercises to help Nancy address her anxiety. Although she conceptualized the anxiety as related to attachment issues (psychodynamic), she also conceptualized Nancy as having cognitive schemas that contributed to her feeling a lack of responsibility and ability to help herself. Given the time constraints related to Nancy's travel plans, the therapist chose to focus on the present aspects of Nancy's problems (while being aware of connections to the past) primarily in the cognitive and behavioral domains. Her thinking was that Nancy was too overwhelmed and cut off from her feelings and that focus on affect and relationships would have to wait until Nancy's return from England.

You may also elect to utilize a strategy from a theoretical perspective or emphasize that approach in your conceptualization even if you do not adhere to that theoretical model. For example, Nancy's therapist utilized a Gestalt double-chair strategy where Nancy imagined herself in one chair and her mother in the other chair and conducted both parts of the dialogue, in preparation for a later mother/daughter session. She was hoping to encourage Nancy to become aware of her mother's perspective in order to foster mutual understanding while at the same time broadening Nancy's constructs of the mother/daughter relationship.

EXERCISE 7.12 ■ What are some particular strategies that you would utilize with Nancy to achieve the treatment goals you designated in Exercise 7.7? Make a list of at least three different strategies that might be selected and consider the pros and cons, the necessary resources and time, and what would be required from you as a clinician and from Nancy as a client. Try to articulate your thinking about each strategy—how and why do you choose it? What are your expectations?

EXERCISE 7.13 ■ How do the interventions you selected relate to your theoretical perspective? How do Nancy's developmental stage, gender, class, race, ethnicity, religion, geographical location, and other variables influence your choice of strategies.

EXERCISE 7.14 ■ What might hamper the effectiveness of your interventions? How will your level of experience factor in and how willing do you think Nancy is to do the work to change? Remember she appears to externalize and blame others for her problems. What are your thoughts about the timing of your interventions?

DIAGNOSIS

The concept of diagnosis comes from the field of medicine whereby the nature and circumstances of a disease are determined by scientific examination so that the appropriate treatment can be applied. As mentioned in Chapter 2, there has been controversy about how relevant this concept is to mental health. There are both benefits and challenges to utilizing diagnoses, for the client and for the therapist.

The classification system of diagnosis can be helpful in facilitating communication and in encouraging therapists to consider the interactions of symptoms and their common co-occurrence. Diagnosis provides a "shorthand" communication between helping professionals, where one can communicate significant information about a client with just a few words because those few words reference a whole body of knowledge that is shared by the professionals. Diagnosis can also help the therapist consider hypotheses about co-occurring symptoms, etiology, and effective strategies for intervention. Finally, some clients experience relief when they realize that their experiences are not just "their" problem, but that others have shared similar experiences.

For some clients, a diagnosis validates that there is a problem, and they are not just "making it up." In fact, a diagnosis can empower a client with information and understanding that can mitigate the effects of stigma and scapegoating, such as the client previously mentioned who indeed had nonverbal learning disabilities. When the meaning of this diagnosis was explained to her husband, he stopped berating her for "being stupid." A diagnosis may also help a client feel that there is hope for change, that others have experienced similar difficulties, and that there *are* guidelines and research that can be used to address the difficulties they are experiencing.

Alternatively, the use of diagnosis can contribute to narrowing conceptualization in ways that contribute to less effective interventions. Diagnoses are descriptions of symptoms decontextualized from the rest of the client's experience. They emphasize pathology (symptoms), rather than strengths of the client, meanings of experience, or functions of behavior, thoughts, or feelings in the client's life. Furthermore, as discussed in Chapter 2, diagnoses reflect a meaning of pathology that is based on expert opinion and normative comparison, both of which are affected by cultural aspects and systems of power and privilege. These issues contribute to controversy about the validity of psychiatric diagnoses and concerns about negative effects of diagnosis.

EXERCISE 7.15 ▪ Discuss the following questions in small groups: where do the diagnoses in the DSM come from? How are they developed and by whom? What criteria are used? What are some diagnoses that have changed or been dropped from the DSM? What do these changes tell you about how diagnoses are created or maintained and their validity? What are some diagnoses that are being currently considered for addition? What are some of the controversies about the inclusion of these diagnoses? What do these controversies tell you about how diagnoses are created or maintained and their validity?

Thus, the use of a diagnosis can be helpful or constraining (or even harmful), depending on the awareness and approach of the therapist. First and foremost, it is vital to understand that a diagnosis is not at all the same as a conceptualization. A diagnosis can be one helpful tool in a conceptualization, but it does not take the place of the contextualized understanding and integration of information we have been discussing. Second, it is important that therapists understand the bases and limitations of diagnoses and actively work to expand their understanding of the client beyond diagnosis—to utilize the potential positives of diagnosis while guarding against the problematic aspects. For example, Dickerson and Zimmerman (1995) explore conceptualizing presenting problems as a) diagnosis, b) functional patterns that come to be understood through dialogue between the therapist and the client, and c) cultural or personal discourses about how people are or should be. They examine how these different approaches to understanding problems affect the types of knowledge that the therapist might emphasize and the related effects on the client, the therapist, and the therapy goals and intervention planning.

While a diagnosis is not necessary for conceptualizing a client or developing a treatment plan, in the vast majority of cases, you will need to assign a diagnosis to your client for billing or organizational reasons. A diagnosis can be a positive thing not only because it may be a necessary thing for the client to receive needed care, but also because the need to assign a diagnosis can raise questions that can be helpful to both the therapist's process of conceptualization and the therapist's goal to avoid the damaging aspects of diagnosis. For example, "medical necessity" is deemed by many third-party payors to be necessary for reimbursement. What does this mean and to whom? How does it affect how clinicians diagnose people who, say, present with symptoms that best fit the diagnosis of adjustment disorders? Do they use a diagnosis that will receive authorization for more sessions? Do they discuss the diagnosis with the client? Do they select the most expedient type of treatment rather than what might be the most appropriate given the client's circumstances? For example, in order to receive insurance reimbursement for sexual reassignment surgery, an individual may have to have a psychiatric diagnosis of gender dysphoria, but that inherently pathologizes the experience and the client may not experience their needs or feelings as pathology.

The ethical issues emerging from the social, political, and economic variables affecting mental health care are a book in itself and are beyond our scope here. However, regardless of whether you think that diagnosis is ultimately a good or a bad thing for the individual, the profession, and our society's view of mental health and mental illness, our current treatment system requires that we affix a diagnostic label for legal and reimbursement purposes. Therefore, it behooves us to think carefully about our purposes, how we can understand the concept of a diagnosis generally, and how we use the diagnostic process with a given client to maximize possible positive effects and to ensure quality care as opposed to harm.

EXERCISE 7.16 ■ Discuss in a small group the aspects of applying diagnosis with which you are comfortable and those with which you are not. How can we be sure that our diagnoses are accurate, given that we do not have definitive tests to establish the presence of most mental and emotional disorders? Consider how much of the concept of diagnosis generally is a social construction. What are the implications of diagnosis for the client? The therapist? The setting? The reimburser? Society in general? How can we use diagnosis as a positive tool to facilitate change?

There have been innumerable studies over the years presenting the same case to different mental health practitioners and resulting in a variety of diagnoses rather than a unified response (see review in Corey, 2009). As previously discussed, we need to consider the individual's stage of development, environmental circumstances, gender, race, ethnicity, sexual orientation, and so forth before arriving at a diagnosis. And we also need to consider that our preliminary diagnosis is likely to be modified as we continue the reciprocal cycle of conceptualization and treatment over the course of our relationship with a client.

EXERCISE 7.17 ■ If the treatment context with Nancy required you to record a tentative DSM IV multi-axial diagnosis at this point, what would you select and why? How would you feel about this? What rationale would support your selection? How could you separate facts from hypotheses?

EXERCISE 7.18 ■ In the DSM IV, read what is written for Adjustment Disorder with Mixed Emotional Mood, General Anxiety Disorder, and Post Traumatic Stress Disorder. Do any of these diagnoses pertain to Nancy at this point? Discuss in small groups and see if you can agree which of these three diagnoses might pertain to Nancy and why.

Nancy's therapist noticed her nervous gestures, which were congruent with the verbal descriptions of her difficulties and self-reported "anxiety," wondered about the interpersonal conflicts with family and peers, and was also aware of Nancy's academic success and focused career direction. She also observed how difficult it was for Nancy to develop trust with her and to be more forthcoming in sessions. Nancy's therapist was required by her agency to record a DSM five-axial diagnosis in writing after the first four sessions. Here are her preliminary thoughts:

- **Axis 1**: Generalized Anxiety Disorder. She considered an adjustment disorder with anxiety, but given the significant persistence of Nancy's anxiety and physical symptoms along with sleeping and concentration difficulties and her experience of Nancy as tense and irritable (particularly noted in the family session), she felt that the GAD diagnosis was more correct.
- **Axis 2**: Deferred. She did not feel she had enough information in these beginning sessions to assign an Axis 2 diagnosis, as she was not sure if what she was sensing about Nancy's dependence, self-centeredness, and entitlement was developmental within her social contexts or related to some type of personality disorder. In addition, the therapist was aware that Axis 2 diagnoses are frequently very stigmatizing to clients, and she was particularly wary of prematurely assigning a diagnosis that might harm Nancy in future contexts.
- **Axis 3**: Here the therapist noted that Nancy did get attention from her parents when she was sick.
- **Axis 4**: There are numerous factors contributing to Nancy's anxieties: interpersonal relationships; academic problems due to attention and concentration difficulties; family stresses; boyfriend relationship.
- **Axis 5**: 60. The Global Assessment Functioning is an overall view of daily performance and functioning along with social, occupational, academic, and interpersonal functioning. It is variable and Nancy's therapist felt that for now this was fair.

EXERCISE 7.19 ■ Discuss within a small group your thoughts about Nancy's therapist's multi-axial diagnosis. What do you agree and disagree with? What would you add or change?

EXERCISE 7.20 ■ If Nancy was 1) a person of color and/or 2) from a lower socioeconomic class, and/or 3) adopted, and/or 4) a first generation immigrant, would your thoughts about her diagnosis change? If so, how would they change? Again, discuss in a small group your thinking and feeling about these circumstances. What new questions might you have?

EXERCISE 7.21 ■ Consider how a multi-axial diagnosis might provide a more complete picture of a client than just focusing on presenting problems or psychopathology (Axis 1 and 2). How does a multi-axial diagnosis expand your understanding of Nancy? How does it strengthen the connection between diagnosis and conceptualization? Also consider the importance of a multi-axial diagnosis for clients who seem to have primarily Axis 2 issues. How might giving only an Axis 2 diagnosis contribute to less complete understanding or effective therapy?

Although most patients present with Axis 1 symptoms, some may present with issues that indicate an Axis 2 diagnosis as well, or that even may seem to be the primary issue. However, it frequently takes much more time to identify the persistent characteristics of an Axis 2 disorder and to have confidence that these characteristics are related to an enduring personality disorder, and not to interacting issues related to the acute Axis 1 problem. For example, self-mutilation may be an indicator of Borderline Personality Disorder, but may also be related to PTSD, depression, or adjustment disorder, and has multiple functions affected by context and relationship history (Suyemoto, 1998). Alternatively, the depression associated with self-mutilation may relate to difficulties with relationships, self-regulation, and other life functioning difficulties that are characteristic of a personality disorder. Listening to clients' stories, assessing clients' insight, and experiencing clients' relational approaches over time enables us to identify the patterns that may indicate that an Axis 2 diagnosis should also be included. However, because Axis 2 diagnoses can have stigmatizing consequences (as noted in Nancy's diagnosis above), they should be particularly carefully evaluated. A multi-axial diagnosis helps us gain a more complete picture of the client and consider possible interacting factors among axes.

FORMALIZING CONCEPTUALIZATIONS AND TREATMENT PLANS

There are many formats that a therapist can use for a treatment plan. One might be to state each problem including the symptoms, complaints, duration, and frequency. You can then translate each problem into goal and objectives statements. Some settings have their own formats and perhaps your instructors also have a model. We think it is important to include minimally the following:

1. Referral Context–how the client came to you and the client's initial expectations and any information from referent.
2. Treatment Context–type of practice you have, private or within organization; possible constraints of the treatment context, such as types of therapy supported, length of treatment allowed, other services available, agency policies, and so forth.
3. Client's presenting problems as client sees them.
4. Related background and contextual information.
5. Your impressions and observations of client.
6. Information from collaterals.
7. Initial conceptualization.
8. Initial diagnosis tied into conceptualization (if required).
9. Agreed upon goals.
10. Initial Treatment Plan: Agreed upon goals and initial strategies.

SUMMARY

In this chapter, we focused on how your understanding of the client translates into diagnosis and treatment planning, using the case of Nancy. We addressed how your understanding of the client is connected to your own orientation and placement on the different dimensions and how this thinking shapes your treatment planning. We also emphasized consideration of the client's sociocultural contexts as well as the treatment context. In the next chapter, we will explore how conceptualization and treatment planning are iterative processes, and continue to follow the case of Nancy as an illustration.

REFERENCES

Ballou, M., (1995) (Ed). *Psychotherapy strategies: Guide to interventions*. Westport, CT: Praeger.

Barlow, D.H. (2010). *Clinical handbook of psychological disorders: A step-by-step treatment manual* (4th ed.). New York, NY: Guilford.

Corey, G. (2009). *Theory and practice of counseling and psychotherapy* (8th ed.). Pacific Grove, CA: Cengage.

Dickerson, V.C., & Zimmerman, J.L. (1995). A constructionist exercise in anti-pathologizing. *Journal of Systemic Therapies* 14, 33–45.

Franklin, M.E. & Foa, E.B. (2007). Cognitive behavioral treatment of obsessive-compulsive disorder. In P.E. Nathan and J.M. Gorman (Eds.), *A guide to treatments that work* (3rd ed.). New York, NY: Oxford University Press. pp. 431–44.

Gallardo, M., & McNeill, B. (Eds.). (2009). *Intersections of multiple identities: A casebook of evidence-based practices with diverse populations*. New York, NY: Routledge.

Lazarus, A.A. (1997). *Brief but comprehensive psychotherapy: The multimodal way.* New York, NY: Springer.

Lazarus, A.A. (2008). *Technical eclecticism and multimodal therapy.* In Lebow, J.L. (Ed.), *Twenty-first century psychotherapies.* Hoboken, NJ: Wiley, pp. 424–452.

Suyemoto, K.L. (1998). The functions of self-mutilation. *Clinical Psychology Review* 18, 531–554.

Watson, H.J., Anderson, R.A., & Rees, C.S. (2010). Evidence-based clinical management of obsessive-compulsive disorder. In R.A. Carlstedt (Ed.); *Handbook of integrative clinical psychology, psychiatry, and behavioral medicine: Perspectives, practices, and research.* New York, NY: Springer Publishing Co., pp. 411–441.

Wolpe, J. (1969). *The practice of behavior therapy.* New York, NY: Pergamon.

8

Iterative Conceptualization and Treatment Planning

As we continue to work with any client, new information unfolds, along with a deepening therapeutic relationship and a sense of collaborative familiarity. Psychotherapy is not necessarily continuous. Each therapy case is different and offers new learning opportunities for both the therapist and the client. Try to visualize the therapy process as a flowing, repetitive cycle; (see Figure 8.1); it is never fixed, always fluid. We are continuously thinking about possibilities, going beyond initial judgments, and avoiding getting stuck in diagnostic ruts. In other words, we are continuing to reassess our story about the client's story based on our developing relationship with the client, new information, and evolving contexts. Thus, conceptualization and treatment planning are continuously adapted and modified.

This chapter focuses on exploring how treatment changes over time in response to new knowledge, which shapes changes in conceptualization and the developing relationship. We will explore the process of reflecting and thinking about the client in between sessions, the feedback loops in the helper/client relationships, and emerging new and deeper levels of understanding. How can the helper develop a continual process of examination of self, conceptualization, and treatment? What questions and strategies aid the helper in avoiding errors and maximizing effectiveness?

We are continually receiving new information about the client, his or her contexts, and understandings of the strengths and challenges that he or she is encountering. New information comes not only from the sources we discussed in Chapter 6 (the client's direct discussion of the issues and contexts,

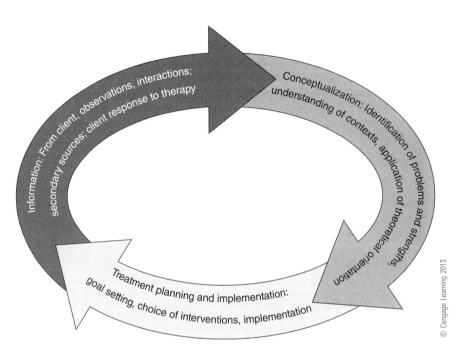

FIGURE 8.1 The cyclical process: Information → conceptualization → treatment planning → response/feedback/information.

observations of the client and the client-therapist interactions in the moment, collateral sources and consultations), but also from the client's responses to the therapy and the interventions unfolding. Everything that we experience with or about a client is some kind of information and is "grist for the mill" of conceptualization. New information leads us to question, refine, revise, and occasionally completely reformulate our conceptualization. A complete reformulation is rare if the initial conceptualization is well-grounded in accurate information and not prematurely developed before the client's presenting issues and contexts have been explored at least in large strokes.

It is difficult to describe the process of continual conceptualization in abstract terms, because it is always so embedded in the unique experiences and circumstances of the particular client. In a basic sense, continual conceptualization means always considering the meaning of new information for our conceptualization, treatment planning and goals, and intervention choices. It means constantly appraising how the client is responding to the therapeutic relationship and interventions. It also means constantly questioning ourselves about how our continual conceptualization is affected by our theoretical orientation and dimensional preferences, to ensure that we are not being led by our own biases, but that we are using our expertise to elucidate the best path to change for the client in his or her particular lived experience. To best illustrate the process, we focus this chapter on two cases: Nancy, who we have been following throughout the book, and Juan, who provides an illustration of some of the different issues that might emerge in continual conceptualization.

CONTINUAL CONCEPTUALIZATION WITH NANCY

The therapist's initial treatment with Nancy focused on her anxiety and her difficulty in managing practical issues (such as financial budgeting and regular attendance at classes). The therapist conceptualized these presenting problems as related to Nancy's feelings of insecure attachment based in early and current family experiences. Furthermore, the therapist conceptualized Nancy's issues with attachment as contributing to a compensatory cognitive schema of entitlement and a lack of a sense of control or responsibility for her own experiences or the reactions she engendered in others. As we have seen, the immediate treatment goals for Nancy were shaped by her decision to study abroad, which would interrupt the therapy process. These immediate goals aimed at developing strategies for Nancy to manage her anxiety in the moment and to manage the practical issues, so that she might have a good experience in her study abroad program.

Nancy successfully engaged in the interventions aimed at meeting these goals, confirming the therapist's conceptualization of Nancy's strengths as including her motivation for change. Nancy learned relaxation exercises for her anxiety that she practiced while in England, implemented a financial strategy to enable her trip, kept a journal of her activities, met her contract with the therapist for attending classes and completing assignments both before and during her trip, and emailed a monthly report to her therapist. Nancy's engagement in therapy and follow-through with homework assignments and self-care were additional sources of information for the therapist's ongoing conceptualization.

While Nancy was away, her brief emails indicated that she was functioning relatively well—attending classes, going out with friends, and taking advantage of opportunities for travel. She reported to the therapist that she kept in contact with her family and, sporadically, with her boyfriend. Nancy expressed a feeling of satisfaction with her success, improvement in her "happiness" and decreased anxiety. One paradoxical, yet common, indicator of the success of these early interventions was that Nancy seemed much less invested in therapy after she returned home at the beginning of the summer. She did not set up regular appointments and saw the therapist only occasionally while she held down a full-time summer job. Nancy seemed primarily interested in discussing her successes and seemed uninterested in engaging in setting new goals for therapy that might address some of the underlying issues that contributed to the pattern of problems she had been encountering. At the same time, she was clearly invested in maintaining some kind of relationship with the therapist.

EXERCISE 8.1 ▪ If you were Nancy's therapist, how would you feel about your sessions with her during the summer? How would you conceptualize her sporadic engagement? How would you interact with her? What, in your own experiences, might affect your thinking about her at this point?

Nancy was going through a "honeymoon" period, where initial therapeutic interventions contribute to positive change. A client may feel that all is well and desire to see their problems as having been much less intense than they thought or feared, and now resolved. In some instances, this honeymoon period "sticks": that is, the presenting problem really is minor or initial changes can catalyze other changes that shift issues underlying more major problems. However, in many instances, the honeymoon period is temporary, as problematic patterns of thinking, feeling, and behaving re-emerge. Even when this occurs however, movement towards change can be facilitated by the client's experience of some change, however temporary or constrained to specific issues. A challenge here is in framing a longer-term process of change as cyclical, supporting the client's moves towards change, and working to help the client (and yourself) avoid seeing the re-emergence of difficulties as a hopeless regression or an indicator that change is not possible.

As would be suggested by the therapist's initial conceptualization of Nancy, which identified more long-term problems, Nancy's honeymoon period was temporary. Four weeks into the fall semester of her junior year, Nancy called for an appointment "in crisis." Nancy said she was feeling overwhelming anxiety, was deeply distressed about her living situation with her suitemates and her relationship with her secret boyfriend, and was having increasing conflict with her parents. She returned to therapy for weekly sessions until Christmas break when she worked long hours in her retail job and was just "too busy to schedule an appointment."

As therapy continues, the distinction between what the client presents as information about the presenting problem and what the client presents as information about his or her contexts frequently becomes less distinct, unless the therapy is specifically structured to examine changes or new information with these distinctions. For example, some cognitive-behavioral approaches will explicitly evaluate changes in the presenting problems (level of anxiety, obsessive thinking, compulsive behaviors, avoidance behaviors, and so forth) and present a structure for explicitly considering what contextual variables relate to those changes. In a less structured or directive approach, the client's understanding of changes in his or her presenting issues is frequently embedded within discussions of his or her contexts and ongoing life experiences.

Nancy was very unhappy with her current living situation, sharing an apartment with two new suitemates she had found through the campus housing office. While in London, Nancy had learned that her suitemates from the fall semester had not included her in their living arrangements for the coming senior year. The only time she saw her previous suitemates was when she called them; they never initiated contact with her.

In relation to her boyfriend, he had—as Nancy had feared—become involved with another woman while she was abroad. Although involved with another woman, he still wanted to "hook up" with Nancy and she was allowing this, as she reported that she needed him to be her friend. She had never been able to make new friends to whom she felt close in college, although she did join groups to "go out" which meant bar-hopping. Nancy felt in general that the other students were unresponsive or hostile to her. She continued to depend on her high school friends who attended college out of town and whom she only saw during semester breaks. She felt that her boyfriend was the only "real" friend she had now in her life who responded to her needs, in spite of the fact that their time together since her return was spent solely having sex in her parents' car, after which he would quickly leave. Nancy did not reveal this until midway through the fall semester.

Throughout the fall semester, the heated arguments with her parents seemed to increase. During her time abroad, Nancy had met her goals of contacting her parents no more than once a week and managing her own finances without requesting additional money from them. During the summer after she returned, she had continued her financial independence through her full-time retail job. However, as the semester progressed, Nancy described how the increased stress from her suitemates and boyfriend made it difficult to meet the demands of her classes and working part-time at the job that she had continued from the summer. She demanded that her parents buy her a car and supplement her spending money so that she might decrease her hours at work, increase the convenience of travel, and, therefore, better concentrate on her studies.

The therapist noticed that Nancy was increasingly agitated. Over the summer, she had felt that Nancy was distant and unengaged, but when the therapy started again in the fall, Nancy was more engaged, speaking quickly, and "gushing" as if she could not describe her distress quickly enough. Nancy also seemed much more emotionally expressive, particularly as the fall progressed. The therapist also observed that she felt much less annoyed at Nancy than she had the previous fall, and she was aware of Nancy's intense pain that at times felt like an almost physical presence in the room.

CONTINUAL CONCEPTUALIZATION AND INTEGRATING NEW INFORMATION

Continual conceptualization is a process of integrating emerging information with information that has already been gathered and organized into an initial conceptualization. One way to approach this is to continuously ask and answer the organizational questions presented in Chapter 6, attending to how new information has changed or modified the initial conceptualization. The other, simultaneous, approach is the development and "testing" of hypotheses based on earlier information and conceptualization. Thus, our continuous and future conceptualizations are inherently related to our initial conceptualization. We do not approach organizing new information with a blank slate, but rather as a puzzle with the picture gradually emerging as new pieces are added. At the same time, we must be open to the possibility that our original picture was so incomplete that a whole new picture may emerge. We then use our revised conceptualization to affect the ongoing development of treatment goals and our choices of strategies.

Revisiting Organizing Information for Continual Conceptualization

EXERCISE 8.2 ▪ Consider the organizational questions from Chapter 7 in relation to Nancy, with the new information we have:

- What is causing the client most distress?
- What do you see as the client's most pressing problems? This might be different than what is distressing to the client.
- What contextual issues seem most relevant? Consider different levels and aspects of context and their interactions.
- How are the client's problems or seemingly problematic ways of thinking, acting, feeling, or relating serving the client?
- What are the client's strengths? What resources are available to the client? Internally, relationally systemically?
- What challenges exist for the client in relation to making change? Internally, relationally, systemically?

How do your answers to these questions relate to your earlier answers? Consider the overall conceptualization you had of Nancy and her contexts. How does your current understanding build upon that conceptualization?

Testing Hypotheses from Prior Information and Conceptualizations

EXERCISE 8.3 ▪ Given your initial conceptualization, what hypotheses might you have had about Nancy's current feelings, thoughts, and behaviors? What kinds of information would help you "test" these hypotheses? How would the process of testing these hypotheses affect how you approached Nancy, the goals you might have set then? With this emerging information, what new hypotheses would you develop and how would you test them?

We are continuously developing new hypotheses and testing them out. And this process influences how we relate to the client, what sort of information we elicit and how. We do not erase earlier conceptualizations and hypotheses; we reassess and revise accordingly.

EXERCISE 8.4 ▪ Given your revised conceptualization and hypotheses, what goals would you set for Nancy at this stage in the therapy? Remember to consider what the client's goals are in relation to your goals and to consider how you would work to connect these. Consider your short-term goals (5–10 sessions, for example), as well as longer-term goals. What particular strategies will you utilize to achieve the treatment goals you have designated?

As a therapy progresses, and as you develop as a therapist, you may find that your conceptualization becomes more embedded in an ongoing "story" about the client, with a foundation in your theoretical orientation. If this becomes a more comfortable approach for you, then asking and answering structured questions such as those above may become less useful. On the other hand, you may find that, even in later stages of a therapy and when you have developed more expertise, it is still helpful to start with questions as prompts for your thinking. Asking questions can also be a check against getting too attached to a particular story or understanding. Sometimes structured questions can help us get a new perspective on something that we might not have explicitly thought about if we had simply been revising a story that we had some attachment to as the "right" story.

As the information above emerged, the therapist modified her conceptualization of Nancy. The therapist wondered about Nancy's lack of communication about the rejections from her previous suitemates and boyfriend. She hypothesized that the pain of rejection was too devastating for Nancy to talk about, and that perhaps Nancy feared that telling the therapist about the rejection would show the therapist how unworthy Nancy was. She wondered whether Nancy feared that this would lead to abandonment by the therapist as well. The therapist began to see more clearly a pattern of how difficult it was for Nancy to trust others, to be vulnerable rather than demanding, and to compromise in order to get along with others. She wondered whether Nancy experienced a distinction between compromise and sacrifice, whether Nancy could imagine a relationship where both parties' needs were met, at least minimally. She continued to understand attachment issues as underlying the problems distressing Nancy. In fact, she felt that attachment issues were even more central than she had previously conceptualized.

The therapist felt an increasing recognition of Nancy's need for relational intimacy, and her fear of abandonment and isolation. The therapist began to reconceptualize Nancy's core cognitive schema from "Things should be the way I want them." and "I should have what I want when I want it." to "If others don't give me what I want, then they don't care for me. And that means my needs are intolerable, but I can't admit that to myself because that would mean that I would be unlovable and all would be hopeless." Thus, she began to conceptualize Nancy's seeming entitlement as more related to a defensive fear of rejection, rather than a lack of concern for others. She also conceptualized Nancy's feelings of a lack of control and agency as related to this cognitive schema, wondering if Nancy felt even more frightened and hopeless by the rejection of others because she did not know how to care for herself.

Given her revised conceptualization, the therapist set the following treatment goals and strategies:

- Continue to develop a strong empathic and supportive relationship with Nancy to address underlying issues of insecure attachment, fear of abandonment, and fear of being interpersonally vulnerable. The therapist continued to validate Nancy's feelings of distress. She also demonstrated that she could tolerate and "hold" Nancy's pain, using this affective relational strategy to challenge Nancy's fear that her needs and emotions were intolerable and unacceptable.
- Another goal was to deepen the trust so that Nancy would become more easily and naturally forthcoming. Because the email check-ins had been so successful during Nancy's time abroad, Nancy was asked to provide an email check-in midweek between sessions.
- Develop Nancy's ability to take the perspective of others and consider how she may contribute to the reactions others may have to her, in order to help her develop more effective interpersonal strategies. Given that one of Nancy's most distressing problems was the situation with her suitemates, the therapist thought about how she could help Nancy achieve a reasonable level of cordiality with them. She utilized role-plays and cognitive restructuring strategies to help Nancy move towards this goal and also developed with her some behavioral plans to engage more with her suitemates.
- Develop Nancy's ability to care for herself and to feel empowered to meet her own needs. As a part of this goal, the therapist worked with Nancy to explore her relationship with her boyfriend. With the help of her therapist, Nancy was beginning to see how her boyfriend was "using" her for sex she really did not like while moving in with his new girlfriend. The therapist utilized assertiveness training to teach Nancy to say "no" and to not always be available when he wanted her. She also used cognitive restructuring to introduce Nancy to more to the concept of reciprocal, mutual relationships. She utilized psychoeducational strategies to teach Nancy about caring sexual relationships. Nancy began to feel better about herself and see how she was being used and how she reinforced this by always being available. The therapist continued to use the relaxation exercises and academic contracting that had worked earlier to help Nancy manage stress and meet her academic goals.

As the fall semester progressed, Nancy continued to be highly engaged in the therapy. She seemed increasingly more comfortable in sessions, and more willing to share her fears and reactions, and to experience difficult emotions in the moment. She continued to have difficulty taking the perspective of her suitemates. Her periods of cordiality with them were intermittent and she had little tolerance for her suitemates' choices or needs. As she began to say "no" to her boyfriend's requests for her to meet with him only for sex, the relationship dwindled. Rather than feeling the panic she had feared from being abandoned, she felt able to recognize that she was taking care of herself by saying "no," and that the ending of this relationship was her choice rather than his abandonment. She felt positively about her ability to "survive" without him. She also felt relief at no longer having what she felt was a shameful secret relationship.

By following Nancy, we are illustrating the cyclical process of information, conceptualization, goal setting, implementation, and response. While we have focused relatively less on the unique person of the therapist in this discussion, it should be clear by now that the therapist's conceptualization, goal setting, and interventions are inherently embedded in the therapist's orientation, dimensional preferences, and skills. If you have a different orientation or dimensional preferences than the therapist who treated Nancy, this may have become evident to you.

In this revised conceptualization and treatment planning, it is still evident that the therapist is influenced by some theoretical orientations and dimensional preferences more than others. The therapist's inclusion of psychodynamic orientation is reflected in her continued emphasis on Nancy's experiences of attachment and on her implicit belief that she (like all human beings) desires to be connected to others, experiencing a "drive" towards relatedness reflecting object relations theory. She also includes a cognitive–constructivist approach, evident in her explicit consideration of the beliefs that Nancy has developed about herself and relationships from her early developmental experiences. Her understanding is also informed by feminist therapy and values, in that she considers Nancy's relationship with her boyfriend in relation to women's socialization that to be single is to be undesirable.

In relation to dimensional preferences, at this point, the therapist continues to focus more on affect and cognition, while also utilizing behavioral strategies such as relaxation to enable experiential enactment of cognitive and affective changes. She also has a more present focused approach, although she conceptualizes the development of Nancy's difficulties as related to her early family relations. The therapist is more active at this point, as well as more structured in relation to process. While still valuing a humanistic orientation, the therapist has moved towards a more active approach where she is seeing the relationship as necessary, but not sufficient. She understands the validation of Nancy's feelings as an important strategy, but is also considering how the relationship may be used to challenge Nancy's fears and beliefs, so that it becomes more of a means to change. The therapist is also acutely aware of the interpersonal process between herself and Nancy, but has chosen not to share process interpretations with the client.

Also, while we have focused less explicitly on Nancy's contexts, the therapist's conceptualization and treatment planning and implementation were developed within her ecological understanding of Nancy. If Nancy had been from different ecological contexts in relation to race, culture, class, sexual orientation, gender, or other variables, the therapist may have made different choices. Similarly, if Nancy's own contexts had meant something different to her, the therapist would likely have had a different conceptualization—for example, if Nancy had been devoutly religious with strong familial values against premarital sex, the meaning of the sexual relationship with her secret boyfriend could have been quite different.

This cyclical process of conceptualization and treatment frequently proceeds in ways similar to what is described above, with information that deepens and expands our understanding and helps us to identify the most helpful short-term goals to work on and to best understand how these fit into larger goals and issues with which the client may be struggling. At other times, new information comes to light that leads us to consider a more major reconceptualization, or to expand our treatment plan to include adjunct treatments.

ENCOUNTERING SURPRISES: THE PROCESS OF ITERATIVE CONCEPTUALIZATION

As the fall semester drew to a close, Nancy seemed to pull back from therapy, stating that the demands of her classes were overwhelming and she needed to focus on her papers and exams. The therapist saw her only once a month in December and over the winter holiday in January, although Nancy kept in contact via weekly emails as she had while abroad. During this time, Nancy again seemed to withdraw from discussing more emotional or difficult topics. In the beginning of the spring semester, however, Nancy returned to regular sessions, again complaining of intense anxiety. Early in the spring semester, Nancy disclosed to the therapist that, in spite of her intense focus on her schoolwork at the end of fall semester, she had received barely passing grades in her classes, about which she was very upset.

The therapist was surprised to hear about Nancy's academic difficulties, given the progress that she had seemed to be making during the fall, particularly in relation to her boyfriend and her report of continuing to successfully use earlier strategies of relaxation and planning to manage her anxiety and daily tasks. She was also surprised because it was another instance where Nancy had not shared anything about the difficulties as they were emerging. In fall sessions, Nancy had focused on her relationships with her suitemates, and her related feelings of rejection and anxiety from the conflict. The therapist wondered whether perhaps she had been too structured, or whether Nancy had felt that she was being directive in content (encouraging the discussion of some things and the exclusion of others) rather than being structured in process. Although these questions arose for the therapist, given Nancy's prior history of not sharing difficulties, the therapist also wondered whether there was some reason underlying both Nancy's academic difficulty and her reluctance to discuss it in therapy. To test these hypotheses, the therapist encouraged Nancy to explore what had been going on that made it difficult for her to succeed academically.

At first, Nancy stated that she had been distracted by socializing, going out with friends on weekends and occasionally during the week. She initially framed this as related to the goals she had been working on in therapy, to develop friends. However, as the therapist questioned how these relationships had been detrimentally affecting Nancy's schoolwork, it became clear that these social occasions and friendships centered almost exclusively on going to bars "to meet guys" and drink. Nancy described drinking more than 10 drinks on each of these occasions. Nancy stated that she drank because she was too "uncomfortable" to talk to people she did not know and that she "panicked" whenever she was approached by someone or when her friends left her side. Even after consuming many drinks, she would panic if approached and would eventually leave the bar.

After a few sessions in the spring, Nancy disclosed to the therapist that she would go home after leaving the bar and "stuff herself with whatever food was around." These episodes of binge eating lasted for hours, sometimes even continuing after she slept and awoke the next day. She would then be filled with shame, self-loathing, and despair and unable to concentrate on her schoolwork or other activities for several days. Nancy disclosed to the therapist that these episodes of binge drinking and eating had been going on even before she had gone abroad, although they had been less frequent. She had not previously discussed them with the therapist.

Nancy insisted that her drinking was not a problem because "everybody does it" and because she got "sick" rather than "drunk." As the pattern of her drinking emerged, Nancy focused strongly on her need to engage socially and her belief that bars and drinking were the only way to do this. She seemed desperate to make connections with others while simultaneously graphically and tearfully describing the intense anxiety she experienced during social interactions. Although Nancy denied that her drinking was a problem, she was concerned about the episodes of binge eating. However, it took some time for her to accept the therapist's observations that the drinking and the eating binges were connected. As the regular pattern of binge eating emerged, the therapist explored with Nancy the possibility of entering an eating disorders program as an adjunct to the primary therapy. At this point midway in the spring semester, Nancy was unwilling to do this.

During the latter part of the spring semester, these episodes of binge drinking and eating escalated. Twice during the semester Nancy reported that she was taken to a hospital emergency room from a bar for alcohol poisoning. As this cycle continued to spiral downward, Nancy stated that she felt more and more "out of control." In a family session, her parents reported losing tolerance after the two midnight calls from the emergency room. They seemed completely disgusted with Nancy and were not looking forward to her returning home for the summer. While the therapist could understand the parents' concerns, she was saddened by their coldness and obvious rejection.

Two weeks before the end of the semester, Nancy reported that her mother had threatened to stop the therapy because the therapist was not on the family's health plan. With this pressure plus the approach of final exams, Nancy was immobilized; she could not attend to her academic assignments and was drinking and binging several times a week. She began to miss classes and feared that she would not pass her courses.

EXERCISE 8.5 ■ Consider the organizational questions from Chapter 7 in relation to Nancy, with the new information we have:

- What is causing the client most distress?
- What do you see as the client's most pressing problems? This might be different than what is distressing to the client.
- What contextual issues seem most relevant? Consider different levels and aspects of context and their interactions.
- How are the client's problems or seemingly problematic ways of thinking, acting, feeling, or relating serving the client?
- What are the client's strengths? What resources are available to the client? Internally, relationally systemically?
- What challenges exist for the client in relation to making change? Internally, relationally, systemically?

EXERCISE 8.6 ■ How do you conceptualize Nancy with this new information? Consider the overall conceptualization you had previously of Nancy, her contexts, and her presenting problems. How does your current understanding build upon your earlier conceptualization and hypotheses about Nancy? Discuss in small groups.

EXERCISE 8.7 ■ Revisit your diagnosis of Nancy. Would you revise or change this diagnosis now?

EXERCISE 8.8 ■ Given your revised conceptualization and hypotheses, what goals would you set for Nancy at this stage in the therapy? Remember to consider what the client's goals are in relation to what your goals might be and to consider how you would work to connect them. What particular strategies will you utilize to achieve the treatment goals you have designated?

EXERCISE 8.9 ■ Binge drinking is currently a major health issue on high school and college campuses. In small groups, discuss your own observations about substance use and abuse in today's youth. As a group, develop a program or plan to deal with this issue in schools and colleges.

When new symptoms or problems emerge, it can be challenging to know how to re-conceptualize the client in context. While sometimes the emergence of new symptoms or problems requires a therapist to step back and reconsider all that he or she has previously thought, it is also important to not get completely caught up in the immediacy of new problems. It is a careful balance between being too attached to one's prior conceptualization and treatment plan and therefore not attending enough to new information and issues and being too reactive to new issues and therefore not seeing the client wholistically in context, with complicated connections between issues and symptoms. It can be particularly easy to fall into a reactive mode when new symptoms seem dangerous or of crisis nature. However, while the danger or crisis obviously needs to be addressed, the conceptualization of the client needs to encompass not only the new symptoms, but also information and patterns that have been previously identified.

As this new information emerged, Nancy's therapist considered how her binge drinking and eating related to her prior conceptualization and the patterns of thoughts, feelings, behaviors, and relationships that Nancy had demonstrated over the past year. The therapist understood these new behaviors as connected to the relational issues that had previously emerged. Nancy's drinking was motivated in large part by her intense social anxiety. In addition, food and drink seemed to be Nancy's way of giving to herself and it was difficult for her to identify the feelings that were associated with the onset of the binging, although she had no difficulty identifying the shame and self loathing that resulted. However, the therapist now conceptualized Nancy's anxiety as particularly socially related, rather than generalized. Given the pattern of issues and progress Nancy had made over the last year, rather seeing the binge drinking and eating as the primary issues, the therapist saw these behaviors as Nancy's attempts to manage her social anxiety and fear, self-soothe, establish a feeling of being cared for, and create a feeling of control. Although the drinking and eating were not successful strategies to meet these goals, the therapist hypothesized that these were relatively desperate attempts by Nancy to meet these needs that she did not have other strategies to meet.

The therapist also developed hypotheses about Nancy's a) choice of when to disclose these patterns and b) the escalation of these behaviors. She wondered if Nancy's lack of earlier disclosure was related to her shame and fear that her behaviors would be intolerable to the therapist and if Nancy could now disclose because the therapist had passed the "test" of being able to tolerate her distress and needs related to the "abandonment" from her former suitemates and boyfriend. She also wondered if, paradoxically, the escalation of these behaviors might be related to Nancy's strength in being motivated to change and her growing trust in the therapist as a person who could help her do so.

During this time, the therapist was very concerned about Nancy's well being in all respects. Because Nancy felt (and acted) increasingly out of control, the therapist felt that it was important to increase the structure and activity in the therapy. She also recognized that if she had been taking a more past-oriented approach, then the current situation might demand a shift to a more present-oriented approach. The therapist also realized that she was Nancy's "lifeline"—the only person who provided ongoing support during and between sessions. She had to work very hard to contain her anger at Nancy's parents for their lack of empathy. She had learned from Nancy that this was a family where everyone "did their own thing"; they never had family meals together or shared family activities. There was little verbal interaction among members except for angry fighting, and it became more and more evident that Nancy was indeed the "scapegoat" of the system.

EXERCISE 8.10 ■ What do you think of the therapist's revised conceptualization? Do you agree or disagree? Why? What issues of orientation and dimensional preferences might be affecting how you reconceptualize Nancy versus how the current therapist did?

The long-term goals for Nancy did not change much, in that the therapist continued to see the establishment of secure, positive relationships and the development of empowerment for self-care as primary goals. However, the immediate 5–10 session goals and interventions through which the therapist worked towards these larger goals changed significantly and the therapist set the following treatment goals and strategies:

- The first major immediate goal was to decrease Nancy's binge drinking and eating. To address this goal, the therapist first consulted with specialists in the field of eating disorders and read some of the current literature on best practices. She then chose to use cognitive–behavioral strategies to work on controlling the binge eating. These strategies included Nancy recording precisely what and when she ate; attempting to identify in writing her thoughts and feelings when she ate; eating only sitting down at a table; attending to her thoughts and feelings when she was in a bar; keeping a daily journal of her thoughts, feelings, and activities, particularly interactions with classmates, suitemates, and family members.

- A second major immediate goal was to decrease the detrimental effect of the binging behaviors on Nancy's academic achievement. The therapist saw the latter part of this goal as particularly important because Nancy felt that academia was the "only" area where she had been truly successful. She used structured activities to help Nancy plan her assignments, particularly at the end of the semester when her anxiety became overwhelming. Nancy was to begin her written assignments at least one week prior to the due date; she was to plan the previous evening where and when she would study the next day. Nancy began to keep a notebook to organize her assignments and daily schedule. Daily checking in via email with her therapist enabled Nancy to complete her assignments and to pass her exams and complete the semester. By focusing on her assignment tasks and breaking them down into manageable units, Nancy was able to achieve some mastery of her anxiety about whether she could complete her work.

- Another immediate goal was to address the issue of the continuing therapy; her mother's threat to stop paying for the therapy created intense panic for Nancy, and she wanted the therapist to tell her parents how necessary the therapy was for her. However, the therapist felt that Nancy wanted to be "rescued" and wanted the therapist to run interference with the parents; she believed that this would contribute to Nancy's belief that she could not care for herself. Nancy was, at first, upset by the therapist's refusal "to take care of her." The therapist responded to this by deliberately engaging Nancy as a collaborator, to foster her emerging belief that she could care for herself. Together, they composed a letter to Nancy's parents, explaining the risk Nancy was at, the progress she was making, and how important it was to have continued care at this time.

Nancy was able to successfully complete her courses and end the semester well. After the semester ended, the therapist and Nancy began to prepare for the summer months, both generally and because the therapist would be out of the country for a month in the summer. The therapist discussed with Nancy how she would manage her drinking when she socialized with her high school friends who would be home for the summer. Nancy was encouraged to tell two people about her difficulties so that she could have some peer support. Nancy was resistant to this, saying that she feared rejection. The therapist noted in session that Nancy's ability to recognize and discuss her fear was itself positive change. Simultaneously, she also encouraged Nancy to test the validity of her fear after carefully considering who she might tell and what the response might be. Nancy finally did tell two friends and was pleasantly surprised by their empathic understanding and support. Taking this risk indicated further progress in behavioral change.

Furthermore, with the therapist's support and insistence, Nancy herself looked into the possibility of an intense outpatient eating disorder program that her therapist had learned about from specialists. This was a huge step for Nancy, to assume this responsibility and to make the necessary arrangements independent of her parents. Nancy found a program that would be supported by her health insurance and made all of the arrangements herself to attend daily four-hour sessions during the month that the therapist would be away. With Nancy's permission, the therapist discussed her situation with the eating disorder program's case manager to ensure a smooth transition and support for Nancy while the therapist was away. While the therapist never received a reply to their letter from Nancy's parents, Nancy reported that they were "ok" with her enrollment in this special program and that they were no longer threatening to end the therapy.

The therapist found herself touched by Nancy's efforts to engage in helpful activities and it seemed to her that some of the "entitlement" beliefs were lessening. Nancy's reaching out to friends, working so hard to arrange specialized treatment, and her more open, relaxed collaboration in the therapy process were significant. Nancy was beginning to get glimmers of how she could care for herself, how some of her own ways of thinking and behaving might be contributing to some of the ways that others were interacting with her, and how she might be able to change these patterns.

EXERCISE 8.11 ▪ How would your goals and treatment plan have differed from that of Nancy's therapist during this phase of therapy? Critique what Nancy's therapist decided to do in the preceding description. Consider the similarities and differences between your conceptualization of Nancy and that of the therapist and how these affect the similarities and differences in your treatment plan. Sometimes, different therapists might conceptualize the client differently, but still have similar goals and may even have similar strategies for intervention. How might this be so?

EXERCISE 8.12 ▪ Do you think that Nancy is making "progress" in her desire for change? What contributes to your evaluation of this? What interventions relate to this progress? How did the therapist's conceptualization contribute to or detract from developing interventions that you think would have been or were helpful for Nancy?

ITERATIVE CONCEPTUALIZATION FROM START TO FINISH: THE CASE OF JUAN

It is sometimes difficult to sort out cultural from personal aspects, and the reasons behind the presenting problem. As we have noted above, while it is rare that an initial conceptualization is completely off target, the amount that an initial conceptualization changes in the iterative process can vary considerably. Let's consider another case with very different dynamics. Remember that we are drastically condensing an overall therapy and that iterative conceptualization is never really "finished."

Initial Information

Information from client from first five sessions

Presenting problem. When Juan, a 25-year old second generation Latino, entered therapy, he stated that he was unable to sleep through the night and was "very nervous" at work. He reported that he was getting only four hours of sleep a night and that he thought he should be getting eight. Juan also reported that, as a result of his insomnia, he was often tired and listless or felt too tired even to go to his job as a shipper in a factory. When he did go to work, his boss nagged him to get things done faster than he felt he could, which made him nervous and caused him to stumble and "make mistakes."

This cycle was increasing, and Juan feared that he might not keep his job. He said that he had been told that he would be put on probation if he missed any more days of work unless he brought in a note from his doctor.

Juan was unsure as to how many days of work he had missed in the past six months. He did acknowledge that his sleeplessness and work difficulties had always plagued him but that these issues had increased in severity and duration during the past year. He reported that he had had the same difficulties in his vocational-technical high school years and that his parents and older brother woke him in the morning so he would not miss the school bus and helped him with his homework.

Referral context. Juan's White European American male therapist, age 47, worked in a community mental health agency which served as an EAP (Employment Assistance Program) for Juan's company. Juan had been referred by his supervisor at work and was allowed 20 sessions. The agency provided short-term counseling from a cognitive–behavioral orientation and was focused on teaching people the life skills they needed to function more effectively in their different contexts. When Juan first appeared for therapy, he brought with him a written authorization form from his supervisor, in which the problems were entered as "excessive absenteeism," "lack of attention to his tasks resulting in mistakes," and "not a team player." Juan had obviously read this and reported that it was "pretty accurate."

Family and social contexts. Juan grew up and currently lived in a small industrial community about 20 miles away from a Midwestern metropolitan area. He was the younger of two boys from a working class family. His father had been a factory maintenance man and his mother had worked as a domestic cleaner. Juan described his father as a "man to admire," because he took care of his family and was close to his wife. His parents were in their late teens when they emigrated from Puerto Rico, and both Juan and his brother were born in the mainland United States. Juan grew up bilingual and had spoken exclusively Spanish with his parents. The family had close ties to kin in Puerto Rico and typically spent their summers there. Juan described fond memories of summers spent at his grandparents' house and described a close relationship with his grandmother.

Juan's older brother, Pedro, was a successful student and had been captain of the football team in the academic high school. Juan had attended a regional vocational high school after experiencing academic difficulties in the ninth grade of the academic high school. He said he was not a good student and that he could never "be as good as Pedro."

Both Juan and his brother Pedro had lived with their parents until Pedro was in his mid-20s, got married, and moved into a nearby apartment with his wife. Juan's parents had recently retired and returned to their hometown in Puerto Rico. Prior to his parents' recent relocation, Juan had continued to live at home, eating dinner with his parents and playing cards with his dad every night. Juan currently lived with Pedro and his family, his wife and his two young sons. Pedro and Juan also worked for the same company, but Pedro worked in a white-collar position while Juan worked in the factory. Juan described feeling anxious that he was letting Pedro down by not being successful at his job.

Juan reported that he had liked high school, particularly socially, and felt accepted by his peers there and was included in going to athletic events and parties. His description of his high school friends made it clear that many of them were also Latino and frequently spoke Spanish together. However, Juan no longer had relationships outside of his immediate family. His high school buddies were all married and working and no longer had time to spend with him. Juan referred to himself as "the stupid one," "a social clod," and inadequate in just about every way. When asked about dating, he replied that he had dated girls in groups in high school but had not dated at all since then. Juan said that he wished he could have friends and feel that he was doing something with his life and "getting somewhere."

Work context. Juan reported that he had held temporary odd jobs after graduating from high school until his brother had eventually helped him obtain his current job three years ago. He said that he felt like "he did not fit in with everybody" and he did not have any buddies at work. When asked whether there were other Latinos working with him, he replied "No, but there were some Black guys." He wondered aloud if his supervisor liked him or perhaps everyone thought he was just there because of his "smarter" brother.

Information from observations of the client and counselor–client interactions Juan was casually dressed, short, and somewhat stocky. He seemed uncomfortable in the first two sessions, sitting turned away from the therapist, making only very quick moments of eye contact, tapping his foot, and usually speaking with hesitation and some stammering. Juan seemed more comfortable with direct questioning and responded well to empathic responsive listening. Over the course of subsequent sessions, Juan seemed to relax more, holding his body in a less tense manner, pausing and speaking more smoothly, and making eye contact while listening. Juan seemed particularly uncomfortable when talking about his relationship with his brother and his dating experiences.

The therapist observed that Juan seemed sad when he talked about his parents returning to Puerto Rico. When asked about his living arrangements with his brother, Juan replied that that was the only place he could live, because "family sticks together even if they are different." His affect was flatter when talking about his brother's family than when talking about his parents. Juan did not ask any questions about the therapist or the process of therapy, and the therapist, therefore, initially thought that he was fairly passive and used to others making decisions for him, (for example, the referral to the EAP or how and where he lived).

Secondary sources of information The only collateral information was the referral sheet from Juan's supervisor. However, the Beck Depression Inventory was routinely administered at this agency and Juan's score indicated moderate depression.

EXERCISE 8.13 ▪ Based on the above information, what would your initial conceptualization, goal setting, and treatment planning be? Discuss in small groups.

EXERCISE 8.14 ▪ What kinds of personal experiences do you have that are similar to Juan's? That are different than Juan's? How much knowledge do you have about the sociocultural aspects that apply to Juan—Puerto Rican culture, bilingual experience, working class experience? How would your own experiences and knowledge (or lack of these) affect your work with him?

Initial Conceptualization and Treatment

The therapist's initial conceptualization focused on the disruption to Juan's family support resulting from his parents' return to Puerto Rico. Although he saw this event as a precipitating factor for Juan's immediate distress, he initially thought that Juan was also generally underdeveloped with regard to interpersonal skills and self-sufficiency. In addition to being generally passive, he saw Juan as overly dependent on his family. The therapist conceptualized that Juan had always relied on his family to steer and monitor his activities and now that his parents were no longer living locally, he seemed to be lost. He thought that Juan's lack of sleeping and relational withdrawal, which he saw as symptoms of depression, might be related to this dependent role in his family. He also wondered about the relation of Juan's depression and anxiety to his envy of his seemingly more competent brother and his expressed yearning for meaning and connection in his own life. The therapist also wondered whether Juan might be struggling with his sense of his own identity, either as a Latino, as a man, or in relation to a general sense of self-worth. He thought that Juan's dependency on his family and his passivity might undermine Juan's sense of himself as a man in the Latino "macho" culture. The working diagnosis at this point was adult adjustment disorder with mixed emotional features.

The therapist's immediate goals were: 1) to improve Juan's sleeping habits; 2) to improve Juan's attendance and performance at work. These were concrete goals that were congruent with the EAP and agency missions and also related to behavioral change that would, the therapist thought, contribute to addressing depression. In addition, the therapist hoped that by establishing a trusting relationship, Juan might begin to open up and explore his thinking processes, self-concept, and feelings about his relationships. The therapist saw himself as having a constructivist and multicultural theoretical orientation in addition to being comfortable with cognitive–behavioral theory and strategies.

EXERCISE 8.15 ▪ What do you think of the therapist's conceptualization? Do you agree? What questions might you have for the therapist if you were talking with him in a case conference? How does your own theoretical orientation and dimensional preferences shape your views of this therapist's conceptualization?

EXERCISE 8.16 ▪ How would your conceptualization, goal setting, and treatment planning be different if Juan was White, Black, Asian American, or Native American? Cuban rather than Puerto Rican? Middle class rather than working class? A woman? Living independently? Involved in an intimate relationship? What impact do you think the gender and race of the therapist might have on working with Juan?

In supervision, this therapist acknowledged that he had been put off by Juan's nervousness during the first session and that this may have affected his feeling that Juan was overly dependent. However, when he had realized how motivated Juan was to "get a life," he felt more positively towards him. As he listened to the therapist discuss his conceptualization, the supervisor applauded his consideration of cultural values and diversity, but cautioned the therapist to consider these possible influences more fully *and* to guard against oversimplification. For example, although the therapist had considered possible cultural influences on Juan's close family relationships, he had not fully considered how these cultural influences might affect interpersonal relationships related to "independence" and achievement. The therapist was encouraged to explore the basis of his own ideas about passivity, dependency, and achievement and his criteria for determining health and mental health. To what extent is Juan's "dependence" on his family healthy or unhealthy? Is this behavior culturally influenced? Is Juan troubled by it and, if so, why? What aspects of his family connections may be more or less likely to be cultural versus indicators of psychological issues or limitations? The supervisor also encouraged the therapist to explore whether Juan experienced his culture as valuing machismo, or if this was primarily the therapist's belief.

After discussions with the supervisor about cultural influences on dependency and achievement, the therapist worked to sort out his own biases. He was now clearer that his positive feelings towards Juan were not primarily related to Juan's acting more in alignment with the European American emphasis on individualism and autonomy, but were instead related to a genuine working alliance towards Juan's goals. In supervision, he also began to more actively sort out how much Juan's living arrangements and "dependency" were cultural, how much they were attributed to Juan's personal style, and whether this style was passive or problematic.

By the end of the first three sessions, as mentioned earlier, Juan and his therapist decided to focus on: 1) sleep hygiene and 2) workplace interpersonal skills, with a longer-term goal of exploring different ways of thinking about himself and his life. Juan's therapist chose an active, directive approach for the first two goals. For example, while exploring Juan's sleep situation, the therapist learned that Juan shared a bedroom with his two little nephews who were restless sleepers. Ear plugs were recommended as well as Juan waiting for the youngsters to be asleep before going to bed himself. This was a problem because the house was small and Juan had no place to be comfortable between dinner and bedtime. So, in the third session Juan and his therapist designed some easy activities outside the home such as taking a walk or running errands for him to try between dinner and bedtime.

In the fourth session, Juan arrived 10 minutes late. He told the therapist that he was "ashamed" because he had obtained earplugs and was sleeping a little more, but he had not been able to initiate any outside activity after dinner. This gave the therapist an opportunity to elicit information about how Juan experiences what he perceives as "failure" and the feeling of shame. It became clear that when Juan was out of his comfort zone and unable to have immediate success, he withdrew, then avoided the task. The shame he felt about this pattern then resulted in his avoiding others who expected him to complete the task, resulting in lateness. The pattern at work was being replicated in therapy.

Because this was the first time this had happened, the therapist carefully questioned Juan about his expectations and interpretation of the assignments. Juan seemed to have felt overwhelmed by the multiple options and experienced feeling pressured to choose one. The therapist was not sure whether these responses and distress were related to cognitive ability, anxiety, or difficulty concentrating that might be related to his sleep difficulties. The therapists and Juan together worked out a detailed and specific plan for Juan on two new assignments: one to address the open time between dinner and bedtime generally and the other to address Juan's social isolation. For the first goal, after exploring Juan's "failure," Juan and his therapist developed a plan of activities for each night of the week, and discussed what Juan would do if something unexpected came up. For the second goal, Juan chose a specific assignment: to ask a workmate if he could go bowling with the guys after dinner one night. They role-played and rehearsed what Juan would say, what would happen if the workmate said "yes," what would happen if he said "no," how Juan might pursue, etc. At the fifth session, Juan came in visibly more relaxed as he reported that he had gone out bowling one night with the guys. In the sixth session, they charted Juan's sleep, checked on his tardiness data, which had improved somewhat, and developed a broader menu for outside activities with considerations for how to choose amongst the different activities.

Juan's therapist operated more in the behavioral dimension at this point in therapy. This is related both to the nature of the short-term goals as well as to the therapist's observations of what approaches made Juan most comfortable. As Juan became more comfortable, the therapist moved more into the cognitive dimension. He began to ask Juan what and how he was thinking when engaging in these new behaviors.

Iterative Conceptualization and Treatment

When the supervisor asked the therapist after the sixth session about his impressions about Juan, he said that he thought Juan was "stuck" not "sick." He reflected that Juan had described how he felt he had not lived up to his perceptions of what his family and teachers expected of him. The therapist noted that Juan's anxiety had become much more acute when his parents left and his routine was disrupted. It seemed that his usual defenses of avoidance and withdrawal kicked in. But when he was threatened with probation from his job and realized that he might become unemployed, he accepted a referral to the EAP and reached out for help. Attending counseling sessions was not common in his cultural community, and reframing this as a mature strength strengthened the relationship.

As Juan's therapy progressed, the charts he kept about his sleeping, tardiness, and new activities indicated some progress on those fronts. However, he did report continued incidents that indicated difficulty focusing, attending, and problem solving. Juan also shared more of his experiences, thoughts, and feelings with the therapist and was more active in discussion rather than just responding to questioning. He confided in the therapist that he had never talked "like this" with anyone before and he really looked forward to his weekly sessions. Juan also began to talk more about his high school social experiences. He described himself as shy, following the crowd, more of an observer than a participant. He never had any particular close friend. He always felt that he had to work harder than everyone else and that he was slower. The therapist wondered if he had underlying attention deficit disorder or other learning disorders and made a note to ask his supervisor if an assessment should be arranged.

Around the thirteenth session, while discussing his high school relationships and his current lack of relationships, the therapist asked Juan when he had had his first sexual experience. The question seemed to take Juan aback. He then revealed that, while he felt he liked girls, he had not seemed to have the same sexual feelings that his friends had. Although he "pretended" to "hit on" girls in high school and felt mildly attracted to them, he felt his friends were much more "obsessed" with sex. While he had had some sexual experiences of kissing and fondling in high school, he stated that he had never been able to sustain an erection long enough to "really do it" and he always wondered if "something was wrong with me." The therapist asked if Juan had any particular thoughts about what might be wrong. Juan responded that he did not think he was a "fag," but just thought something was "missing" in him. He seemed to be relieved to be talking about this but after a short time, he changed the topic back to his increasing discomfort living with his brother. His discomfort living with his brother created dissonance for Juan: on the one hand, he felt obligated to be grateful to his brother and believed that family should stay together. On the other hand, he wanted to get away from his brother's family, but did not know if that was okay or how his parents would see it, and he felt guilty about these feelings.

The information that emerged in this session made the therapist step back and consider again his current conceptualization of Juan. His initial conceptualization of Juan's difficulties as at least partly related to his family relations seemed to still be relevant. He began to be clearer about cultural and personal influences and Juan's own desires, given that Juan had been increasingly expressing ambivalence about his living situation with his brother, and a desire to be more on his own and develop new friendships.

The information about Juan's dating and sexual history brought up additional thoughts about Juan's relations with his family, as well as new considerations about Juan's self-concept that may have been affecting his mood. The therapist wondered whether Juan's early experiences led him to see himself as unable to have (or uninterested in having) intimate relationships with girls/women. He wondered whether Juan's dependence on his family was a way of avoiding the risk of social contacts. As a relationship with a spouse seemed to be the usual catalyst in Juan's experience for living separately from family, was Juan's ambivalence about his family attachment related to his feelings or fears about intimate relationships?

The therapist also wondered about Juan's comment about not being a "fag." He was concerned in a general way about Juan's homophobia and realized that he needed to reflect on how his own values about sexual orientation may be similar or different than Juan's. He was also concerned in a more specific way, because he was not sure whether Juan's self-reported lack of sexual feelings may actually be related to not being attracted to women and, in fact, being gay and struggling with internalized homophobia. Finally, prior discussions with Juan had indicated that Juan did feel that machismo was valued in his family and community, and that he frequently did not live up to this value (as he felt he failed in so many other areas of achievement as well). Given that this was so, the therapist wondered about the effect on Juan's self-esteem of feeling different from his peers and male family members with regard to his sexuality and whether these feelings of failure and alienation could result in his increased retreat into his family.

In supervision, the therapist suggested that maybe an assessment for Attention Deficit Disorder was not the current priority, given this new material. The therapist and supervisor also discussed together the possible meanings of Juan's described lack of sexual interest. Together they discussed strategies for exploring Juan's sexuality. The therapist realized that if part of Juan's lack of sexual feelings towards women was related to suppressed (or repressed) sexual feelings towards men, then it would be necessary to educate himself (the therapist) further about working with sexual minorities in the coming out process. While he did not want to contribute to or collude in Juan's internalized homophobia, he also wanted to respect Juan's own process of coming out or becoming comfortable with his own sexual identity. If, on the other hand, Juan was really not attracted to men at all as he said, then what should the therapist do about Juan's homophobia? While the homophobia would not be central to Juan's own issues, the therapist felt that to not mention it would be to passively accept it and, therefore, collude in something that he felt was socially oppressive. The supervisor and therapist explored ways that the therapist could gently confront or challenge Juan's homophobia without making it the central issue and detracting from the primary goals of Juan's presenting problems.

In supervision, the supervisor and therapist were considering various hypotheses to consider while re-conceptualizing Juan, given emerging information. In addition, the supervisor suggested a medical consultation to see if there was any medical aspect to Juan's reported low sexual desire and impotence. The therapist had not thought of this, but instead had conceptualized Juan's expressed feelings and experiences as related to more dynamic and relational aspects. But he agreed that a physical examination might be helpful, if only to rule out physical issues.

When he suggested the medical consultation to Juan, Juan expressed surprise but said that if the therapist thought it was a good idea, he would agree, indicating that trust had developed. Juan had rarely seen a doctor except when he had his tonsils removed in childhood. The agency had medical consultants available, and it was not difficult to schedule an appointment for Juan between the next two sessions. The results came back indicating low testosterone and some additional hormonal imbalances, which were amenable to medical intervention. At this point, the therapist realized that he needed to continue to reassess his original conceptualization and treatment planning.

Juan was vastly relieved when the medical results were explained to him by both the physician and then the therapist. He felt that the hormonal imbalances were "the answer" to why he was not sexually interested. The therapist's view was that it was possible that this was true, but it was also possible that there were other contributing factors and was therefore cautious in endorsing Juan's conviction that he was obviously heterosexual and "just" had a small medical problem. Juan was able to eventually express that he had spent years hiding, fearing that he was gay, which in his family and culture was a taboo. These discussions emphasized, to the therapist, that there would likely be psychological patterns and issues to be worked through, as it seemed that Juan's internalized homophobia and his tendency to retreat within himself contributed to his inconsistent and inattentive performance at work and social isolation.

In addition, this was an opening for the therapist to explore and gently challenge Juan's homophobia. While recognizing that Juan felt that it was a taboo in his culture, the therapist mentioned that he, personally, felt that it was fine to be gay, although sometimes socially difficult. And that he knew Latino and Puerto Rican people who were gay and quite happy. In doing this, the therapist sought to address the social issue of homophobia being inherently problematic (for anyone) as well as the possibility that Juan may, now or in the future, have sexual feelings towards other men.

The therapist was able to obtain approval for five more sessions from the insurance company because of Juan's new medical diagnosis. At this time, the therapist and Juan revised the goals and treatment plan. Juan expressed a desire to improve his social relationships, particularly related to dating. This seemed related to his no longer being fearful that dating would lead to him being "found out." The last sessions of therapy therefore focused on teaching Juan social and sexual relationship skills and assigning him specific steps in beginning to date. These sessions also processed Juan's feelings and thinking about these experiences, and gradually the assignments required more risk taking on Juan's part. In addition, the therapist suggested that Juan enroll in a co-ed social skills group held weekly in another program at the agency. His thinking was that if Juan began this group while still in individual therapy, he could continue it after the individual therapy ended.

As the hormone treatment began to work, Juan began to sleep better and to have fewer incidents at work of non-attentiveness. He began to relate differently with his co-workers and they even asked him to join their bowling league. As Juan became more visible in the community, he re-engaged with some of his high school buddies who seemed interested in his joining them for their high school's athletic events. He also went on one date with a woman he had met in church and asked out as part of his homework assignment for therapy. Although Juan felt they did not really connect, he felt positive about the overall experience.

EXERCISE 8.17 ▪ Discuss your reactions to these changed goals and treatment for Juan. What would you do differently? How comfortable would you be talking about sex so explicitly with a client of the same gender? Of a different gender? Of a different sexual orientation? With a different cultural background?

The therapist spread out the final three sessions by meeting every two weeks. He realized that he probably would not achieve Juan's goal of independent living separate from his brother in the time they had together, but he worked with Juan to set up a specific plan for how to go about this when Juan felt he was ready. He saw Juan as: (a) more confident: he smiled more, his speech and his body language were more relaxed; (b) more self aware: he was able to laugh at himself and was gentle and empathic as he described the people with whom he was socializing; and (c) more social: he was out more frequently rather than spending all of his time at home. He also seemed less dependent on the therapist. The therapist felt that progress had been made on all of the goals they had agreed upon.

The above case is an example of how midway through a therapy, the need for outside consultation can emerge and new information requires reassessment of original conceptualization and intervention. It is also an example of the importance of supervision and constant self-reflection in ensuring that our case conceptualization best reflects the client's experiences with sensitivity to his or her unique contexts.

ITERATIVE CONCEPTUALIZATION AND MODIFYING DIAGNOSES

As one's conceptualization changes, the initial diagnosis may also need to change. With Juan, the Axis 1 diagnosis stayed the same, but the Axis 2 possibility of dependent personality was no longer a consideration and the hormonal imbalances were an important part of the Axis 3 diagnosis. With Nancy, the

Axis 1 diagnosis changed to social anxiety as it became apparent that that underlay her interpersonal conflicts and drinking and binging behaviors.

Often, trainees who are assigned cases that have been treated by prior trainees report that they do not agree with a prior diagnosis. Not only are they seeing and experiencing the client through different eyes and filters, but also they may see differing contexts. For example, one trainee reported that an adult patient diagnosed with Borderline Personality Disorder reported relationship difficulties in a way that caused the trainee to wonder about Asperger's Syndrome as a primary diagnosis. This patient functioned very well in her work situation (she was a laboratory technician), but had difficulty with family and social interpersonal relationships. It seemed to others that "she just didn't get it" and, in fact, she did not. A subsequent neuropsychological examination confirmed the Asperger's diagnosis. The symptoms of the two diagnoses seemed similar, but the prognoses and appropriate interventions differed drastically. The treatment with the new therapist after the neuropsychological examination then focused more on teaching empathy and social skills. Interestingly, when the diagnosis became less pejorative, her family also treated her differently. While she was not "cured," she worked closely with a caring therapist and her relationships improved considerably.

ITERATIVE CONCEPTUALIZATION, CONSULTATION, AND ADJUNCT TREATMENTS

We saw with Nancy that specialized eating disorder treatment in addition to the individual therapy was essential. Nancy's therapist realized that high level intense treatment of eating disorders was not her area of competence, and she recognized that these behaviors were out of control. We saw with Juan how important it is to not assume that the cause of symptoms is always psychological; medical consultation is an important part of differential diagnosis—we want to rule out or in all possibilities.

Sometimes, clients also assume that their symptoms are psychological. One client, a cancer survivor being treated in psychotherapy for anxiety, was also being treated with a regimen of medications to wipe out estrogen in her system. She never mentioned anything about her sleep difficulties to her therapist or physician until her therapist asked her about her sleep after noting how fatigued she looked. She reported that, due to her anxiety, she woke up every night and vomited. Her therapist responded that this sounded pretty extreme and directed her to call her physician at once. It turned out that this woman was reacting to her medications, which obviously required some change. She had believed that her nighttime vomiting was just another symptom of anxiety.

We suggest that you have your own list of consultants who can help you decide when and how to seek consultation. Many third-party payors insist that a psychopharmacological consultation be sought prior to authorization for more psychotherapy sessions. If that is the case, and you choose to remain on those types of panels, it is imperative that you have some psychiatric consultants who will do objective patient evaluations and discuss with you the advisability of medication. Medication can be very helpful if prescribed and monitored competently, reducing symptoms of anxiety and depression to the point where more effective psychotherapy can occur.

ONGOING RECORDS AND CLINICAL NOTES

The format and requirements for client records vary among settings. Typically, the minimal information required is date and time of service, summary of goals and outcome for a particular session, and any significant changes to initial diagnosis and treatment plan. Many therapists maintain their own clinical notes in which they record their impressions, thoughts, and issues they may want to pursue in the next session. While both could be subpoenaed in a court case, these notes are typically for the clinician's own continuous conceptualization and treatment planning purposes, a way to remind oneself of hypotheses and discrepancies and other more random thoughts and feelings about the client in the particular session.

There is room for debate over whether the clinician's progress notes belong to the client, the therapist, or the agency. Certainly, agency forms that the patient completes belong to the agency although in most states the client is allowed legally to have access to any information the agency maintains. The client may view the therapist's notes as well, so it behooves clinicians to take care with the language they use in note taking. The HIPAA privacy rule protects any notes from being submitted to anyone without the written authorization of the patient, unless there is a legal action and a subpoena. The protected records include medications and prescriptions, the types and frequencies of treatment as well as the start and end of each counseling session, the results of clinical tests, a summary of the diagnosis, functional status, treatment plan, symptoms and progress to date (HIPAA, 2002).

We recommend that clinicians take a few minutes to jot down their own thoughts (or process notes) as soon after the session as possible, while the material is fresh in mind. These notes should be reviewed prior to the next session. However, we find ourselves thinking about clients in between sessions and at any time we may want to add another thought or idea to our clinical notes. In between sessions, we may also want to discuss the client with a supervisor or we may receive new information from a collateral contact (of course, with the permission of the client). Just as we expect clients to "work" in between sessions and put their newly developing self awareness into practice, so do we.

EXERCISE 8.18 ■ In small groups, make a list of the kinds of questions you might ask yourself at the end of a session with a client. How do you assess the outcome of each session? How do you decide where to go from here? The purpose of this exercise is to begin to think about how you not only assess and review what happened in the session, what you learned about yourself, the client and the relevant contexts, but also how you might want to review your conceptualization and treatment plan.

INTERACTIONS WITH SYSTEMS OF CARE

The case of Juan illustrates the impact of the treatment context on the therapy process. His EAP provided short-term focused work on the symptoms affecting his work performance. So the therapist utilized cognitive–behavioral strategies to attend to the behaviors causing difficulties. But as he more deeply explored Juan's functioning in his family, work, and community contexts, he was able to discern some patterns of avoidance and retreat. By developing a trusting relationship, he was able to uncover Juan's secret fears and shame about his sexuality and masculinity. After the medical consultation, Juan was given five more sessions to be spread over several months for follow-up and support.

Many third-party payors consider "medical necessity" as the basis for authorizing further sessions. While recent legislation regarding parity of treatment for mental and physical illness has alleviated this split somewhat, it is important for therapists to advocate for their clients particularly by seeking essential services from other health care systems.

Nancy's family had originally agreed to a therapist who was not on the family health plan. When Nancy began to improve, however, her family brought up the money issue and insisted on a therapist on their list. It was fortunate that Nancy and her therapist were able to act together and that issue did not arise again through the course of Nancy's treatment. Nancy's therapist sought consultation from eating disorder specialists and was able to provide Nancy with some resources for adjunctive specialized treatment.

SUMMARY

This chapter focused on integrating new information for continual conceptualization, the feedback loops in the therapist/client relationships, new and deeper levels of understanding, and reflecting and thinking about cases in between sessions. Using the cases of Nancy and Juan, we explored how new information activates the circular loop of reassessing earlier conceptualizations, revising goals and treatment plans.

We are continuously testing our hypotheses of what information is needed and expanding the contexts of presenting problems. While we may focus specifically on the presenting symptoms, whether or not we have time to explore underlying issues, our broader hypotheses shape how we think about a client and help us to differentiate between what we can provide immediately to meet the criteria of the client and treatment contexts and what we might recommend the client consider for future treatment.

REFERENCE

HIPAA Final Privacy Rule 45 CFR Section 164.501. http://www.hhs.gov/ocr/hipaa/privrule.pdf

9

Continuous Evaluation and Termination

Throughout this book, we have stressed that case conceptualization is continuous and that we are always weighing our initial thoughts and ideas with new information that requires us to reassess goals and treatment plans. In this chapter, we will consider the process of evaluation and its relation to continual conceptualization and treatment planning. We will also consider the evaluation of outcomes and the relation to decisions about termination. Finally, we will consider the process of termination, including transfer of clients and return of clients after time.

We have seen throughout this book that the therapeutic process is fluid and dynamic, circular rather than linear. Human beings are always evolving and contexts and circumstances always changing. Thus, as we develop personally and professionally, gaining new experiences and self-awareness, the ways we conceptualize evolve. Many therapists in midlife say, "I wish I had known this twenty years ago when I was treating…." This does not mean that you cannot be a good novice therapist, but simply that we all change and grow with experience. Clients' expectations for and attitudes about psychotherapy, as well as about themselves and their perspectives, may be different today than they were in previous therapies or contexts.

Therefore, we always need to keep abreast of individual, family, and societal developmental challenges and attitudes. For example, the mores and attitudes about contemporary families are changing: there are an increasing number of gay/lesbian families; single parents by choice; foster and adoptive families; interfaith, interracial, and intercultural families. We need to be able to provide ethical treatment to a greater diversity of clients and also need to continuously expand our own ways of viewing and accepting differences and change. Likewise, we need to keep up with professional literature and research findings and be open to new ways of conceptualizing and providing treatment. Many later career clinicians are now utilizing conceptual and treatment approaches that were not available when they were in training.

CONTINUOUS EVALUATION OF CLIENTS AND THEIR PROGRESS IN THERAPY

Each time we see the client, we are considering how they are thinking, feeling, and responding to us and to the process of the therapy. We are, as discussed in previous chapters, considering new information that emerges. As trust develops, the client reveals more of his or her innermost thoughts and feelings and often shares more pivotal events in his/her life. We liken it to peeling the layers of an onion so that we get closer to the core and can help the clients improve their awareness and choose more effective responses.

We are also continuously evaluating the client's response to therapy. Is the client engaging in therapy and in the change process? Is the therapeutic relationship evolving appropriately? Does the client follow through with plans made in therapy? Is he or she taking more responsibility for choices and behaviors? We are also evaluating the client's response and the effectiveness of our conceptualization and treatment choices. How is the client responding to the interventions? Are our hypotheses about effects and choices of interventions being confirmed or disconfirmed? Do some interventions seem to be more effective than others? Is there evidence that changes from therapy are affecting the client's life more generally? For example, are distressing symptoms abating? Is the client describing applying new understandings or skills? Are the client's relationships changing outside of therapy as well as in therapy?

In other words, part of our ongoing reconceptualization is evaluating whether there is movement towards achieving agreed upon goals. As new information comes to light about the client and the client's response to therapy, we re-evaluate previous conceptualizations of treatment and plan accordingly. When there are time constraints with regard to number of sessions, it is even more important to keep the end in mind as we evaluate how much we can achieve before the designated termination. As discussed in previous chapters, unexpected circumstances can arise in a number of ways but even without surprises, we are responsible for moving along at a pace that is helpful to the client while adhering to the time and resources available.

One way for us to keep abreast about what is happening within the client's life is to check in or ask directly. Many therapists begin each session with open-ended questions such as "How has this week been for you?" "What's happened in your life since I last saw you that I should know?" Periodically, we need to be sure to check in about all of the client's contexts, for example,

"How are things at home? At work? At school?" This enables us to actively incorporate new information and experiences and ensure that our interventions are responding to new circumstances in the client's contexts, as well as changes in his or her thinking, feeling, and behavior. Specific questions such as "What happened when you asked your boss if you could change your schedule?" can indicate that you remember the details that you have previously discussed with the client and have been thinking about the client in between sessions. These kinds of more detailed questions can also act as a springboard to exploring changes in contexts more generally. Thus, checking in about client experiences and contexts is a continuous way of evaluating what is going on with the client.

Similarly, many therapists incorporate clients' own evaluations of progress through checking in or asking direct questions about how clients are evaluating the therapy and the change process. For example, you might ask questions such as: "How do you feel about what we talked about last week?" "What thoughts and feelings did you have after our last session?" "How do you think we are doing? Is this working for you? Is there anything you'd like to change about what we are doing?"

Continuous Evaluation of Our Effectiveness as Therapists

While clients' direct feedback about progress is certainly important, the therapist also has a responsibility to incorporate continuous evaluation into his or her conceptualization of the client and the therapy. Of course, *how* to do this evaluation is the difficult question. Evaluating progress towards agreed upon goals for changes in feelings, thoughts, behavior, and relationships is one part of evaluation. But this kind of outcome progress frequently takes place over time, and evaluation of therapy is frequently framed around the central question of whether the therapy as a whole (several weeks, months, or even years) creates change. However, it is clearly not enough to evaluate effectiveness only when one is considering termination! We need to continually evaluate progress so we can provide better services throughout.

Novice therapists, in particular, are always asking themselves if they are being effective and how they will know. More experienced therapists also ask themselves this question but have the benefit of having treated many more clients, which provides intuitive and evidence based experience to guide their evaluation process. Interestingly, there is little research about how psychotherapists evaluate their own effectiveness. Most of the research focuses on evaluating client outcomes, assuming that positive outcomes are a direct result of the psychotherapy process. Positive outcomes may be related to the therapy process and particular interventions, or to other factors, such as the motivation and timing of the client's decision to change, external reinforcements, and other contextual variables. Could the therapist have achieved the therapeutic goals faster or in a better way? Could another therapist have been more effective with the same patient? The point is that we need to pay attention to how we evaluate our effectiveness as therapist with each client.

Your own feelings and thoughts will guide your self-evaluation as a therapist. It is often helpful to talk to others involved in the case or with peer supervisors to process your self-evaluation. But more importantly, you need some criteria to utilize in assessing your effectiveness in addition to a client's self-report. Such criteria might include observed changes in behavior, thoughts, or feelings; changes in reported activities; information about changes in relationships or behavior from reports from others such as family or school, and so forth. Other criteria may relate to the client's engagement in therapy, such as adherence to treatment schedule and assignments. The particular criteria you use will be related to your theoretical orientation, conceptualization, and goals for the therapy. For example, therapists with a structured theoretical orientation may review written assignments at the beginning of each session in order to ascertain if cognitive and behavioral changes are evident. Specific data collection and analysis will determine whether the client is ready to proceed to the next step of attaining desired changes. Alternatively, therapists with a more psychodynamic approach may assess changes in the relational interactions between client and therapist or the extent of client insight into behavior, thoughts, and feelings.

Evaluating Client Progress

First, we need to specify what outcomes we are evaluating. Are we evaluating the working alliance? The effectiveness of selected interventions? Our progress from beginning to end of treatment? Specific goal achievement? Symptom removal? And, as discussed in the previous section, are we talking about subjective or objective measures of outcomes?

EXERCISE 9.1 ▪ If you have ever been a client, consider how you decided what was helpful and what was not helpful. How did you decide whether your goals and objectives were met? On a scale of 1 to 5 with 5 being high, how would you rank the outcomes of your experience as a client? Discuss the criteria for this ranking. If you have never been a client, apply this to being a student. What was and was not helpful in your learning, particularly in classes that aim to teach more than content, such as practicum classes or therapy supervision? How do you evaluate your own learning?

EXERCISE 9.2 ▪ Discuss in small groups your thinking about how you assess a client's progress and outcomes. What sources of information do you seek? What kinds of questions do you ask yourself? What elements of your theoretical orientation and preferred dimensions shape your questions and thinking?

When thinking about evaluation of progress, it can be helpful to think about assessing different time components of therapy. Evaluation can be assessed in relation to what is happening during a session, between sessions, and over time. Assessing outcomes is a collaborative process involving both the client and the therapist and, sometimes, others such as family, teachers, and other providers. A collaborative therapeutic relationship enables you to check in with the client, focusing on both the client's experiencing of the therapy and how it is affecting the client's life functioning. There is no way, however, to remove subjectivity from this assessment, no matter how many instruments you use to "objectively" measure changes.

Assessing during a session When a client walks in your door for each session, you are immediately reading her or his nonverbal communication such as facial expressions or body posture to assess their mood and wellbeing. You can sense if they are relaxed, tense, upbeat, sad, upset. You may check out your observation and interpretation of this observation by checking with the client, such as "You seem upset/tired/happy/sad today." Or you may verbally or nonverbally encourage them to tell you how they are doing or what is going on. As the session proceeds, you are continuously checking your own feelings and thoughts, the relationship and what type of movement is or is not occurring. As in any relationship, within the therapy hour you certainly can tell from the client's verbal and nonverbal responses if you are connecting and if you are both engaged. But you also want to think about whether you are being effective not just in the moment, but also in relation to the therapeutic goals and to the client's world outside of therapy.

Prior to a session, you will likely have some ideas about topics you want to cover or continue from preceding sessions, related to your conceptualization and agreed upon goals. Depending on your initial observations as the client enters the room or what comes up in a session, you need to be able to go with the client's agenda, not yours. Often we find that in a particular session, we switch focus midway or at the beginning from what we thought we would cover, or we digress temporarily to address a new emerging issue for the client before returning to our planned focus. This "match" and your ability to meet the client where he or she is in a given moment will likely affect your assessment of your effectiveness within a particular session. If a client has experienced a crisis between sessions and wants to discuss this with you, but you insist (explicitly or implicitly through your structure of the session) on focusing on the topics or strategies you had planned, it is likely that the client will be unable to engage as you had hoped. Your evaluation of the session effectiveness should then pick up on the break in the working alliance, and you will likely feel that this session was not as effective as it could have been in helping the client. Remember, because therapy is a circular process, there is always space to return to or revisit "unfinished business" from earlier sessions.

This is not to say that you should allow the therapy to be "derailed." Remember the case of Don from Chapter 6? He was the client who came to therapy to work on intimacy issues, but spent most of his time discussing schoolwork. The difference between the case of Don and the point we are making here is that this was a pattern established over time, related to the therapist's conceptualization. This contrast emphasizes the need to keep your conceptualization in mind, even as you may change the direction of a particular session.

Evaluation of a given session, even if the content or structure is different than you anticipated, should be related to your conceptualization: How does what you did with the client in this particular session relate to the conceptual themes and goals? In addition, in evaluating a given session, it can be helpful to return to your dimensional preferences and their application to this particular client that should be evident in your conceptualization. If you feel that a structured, non-directive, active approach would be most helpful with this client, was this session structured, non-directive, and active? How did the client respond to these particular aspects? If your conceptualization focuses on psychodynamic and existential issues that highlight the role of relational process, how did you manage the process (whether you explicitly interpreted it or not)? It can be challenging to communicate the process of evaluating progress in sessions to novice therapists because this process of evaluation is so strongly tied to the complex process of continual conceptualization and related theoretical orientation. This is another reason why we have emphasized the need to have a clear conceptualization, including an understanding of the development of the problems and strengths, and clear treatment planning, reflecting awareness of dimensional preferences and application.

EXERCISE 9.3 ■ If you have ever been a client, can you remember a specific session that you thought was particularly helpful? What was helpful about it? Did you know at the time that this session would be particularly helpful? Can you remember a specific session that you experienced as less helpful? What contributed to this? How does a particular session relate to your evaluation of the therapy overall? If you have never been a client, apply this to being a student. How do you evaluate whether a particular class session is a strong contribution to your overall learning?

EXERCISE 9.4 ■ Discuss in small groups your thinking about how you assess a client's progress in a particular session. What would be an indicator *in a given moment* of positive progress? What would be an indicator of challenges to progress? How do these relate to conceptualization and theoretical orientation? To your dimensional preferences?

Assessing effectiveness within a session can also be challenging for novice therapists because effectiveness does not always mean feeling good, for the client *or* the therapist. A client or a therapist may feel uncomfortable or distressed in a specific session and the session can still have been very effective. Obviously, there are some clients who will be more of a challenge to our feelings of efficacy than others. This may relate to our discussion about the fit between clinician and client in Chapter 5, to our dimensional preferences, or even to issues related to clients who present with contextual aspects that arouse our own feelings of ignorance or inadequacy. For example, if a client's anger management problems arouse your own uncomfortable feelings about anger, you may find yourself dreading sessions or avoiding dealing with this issue. Such a situation can be an opportunity to stretch yourself out of your usual comfort zone and to broaden your knowledge and skill bases with consultation and supervision. In other words, this is an opportunity for you to acknowledge and work through your own issues that a client evokes. Similarly, when clients present with issues with which you are not familiar or comfortable, this is a learning opportunity to seek out the many professional development opportunities to learn new interventions and treat different types of clients and problems.

Another issue that may arise within specific sessions is the "door knob" phenomenon: when the client drops some critical information as the session is ending. It is important for the clinician to assess whether this is the client's attempt to gain more time and attention, an avoidance of difficult issues by bringing them up when there is not time to explore them, a blurting out of difficult information that evokes heightened shame and anxiety, or something else. In some cases, it may be possible to arrange an additional session or to suggest that this be the first thing to discuss next time. If this is habitual, it is incumbent on the therapist to alert the client to this pattern and try to figure out how important material can be revealed earlier in the session.

There may be times when the client is not happy with something you do or say, such as when you confront him or her about discrepancies or challenge certain beliefs. Although a client's discomfort does not inherently mean the therapist is doing something wrong, it is something that the therapist should seriously attend to and carefully evaluate in relation to the conceptualization, the goals, and the process of therapy. Sometimes, responses of discomfort indicate that we do need to change our approach, carefully consider our conceptualization, and explore the match between our own style and the client's needs. For example, there may be times when you think that you have been too directive or challenging and wonder if the client will return. In reflecting on the client's responses and considering the conceptualization, you may realize that the client needs a gentler, less directive approach in dealing with certain issues. However, this reflection may also lead you to "wait and see," to gather more information about how the client responded and what he or she learned from this approach. The therapist may find out at the next session that the client really took it in and benefited. There will also be times when you question your overall effectiveness with a particular client only to learn later from a subsequent therapist how much that particular client really appreciated his or her work with you.

Experiencing client discomfort or even criticism of the therapist because of this discomfort can be particularly challenging for novice therapists. Novice therapists may feel this because of their own desires to please, their discomfort with others' discomfort, or a belief that effectiveness means "making the client feel better" at all moments, particularly those shared in the therapy room. Although we all have needs to be liked and appreciated, our goal as a therapist is to help the client, not to be liked by him or her all the time. We continuously need to consider whether we are engaging the goal of helping our clients to grow responsibly or whether we are enabling them to stay stuck because they make us feel good.

For example, it may feel good to Nancy to have the therapist's support and attention. Nancy's positive response and appreciation for the therapist at these moments may make the therapist feel effective and positive. On the other hand, Nancy may feel discomfort when the therapist encourages greater perspective taking and empathy for others. At these times, Nancy may show her discomfort or even criticize the therapist. Although the therapist may be tempted to avoid interventions focused on the second goal, both aspects are necessary for the positive change Nancy was seeking. In sum, the client's reactions or responses, particularly in one moment within a session, are not always accurate indications of effectiveness; sometimes, the most effective intervention is uncomfortable for the client.

Of course, we need to evaluate client's responses of discomfort or criticism carefully to ensure that we are not using our power poorly, pathologizing client's responses, or being defensive. In addition, it is important to evaluate whether the client is able to tolerate the discomfort and ready to use the new learning, insights, or emotional experiences. Whether you and the client can navigate these difficult feelings and responses also frequently depends on the timing and context of interventions which create discomfort. Change can be hard and frequently is uncomfortable, but within a strong working alliance clients will be able to tolerate this discomfort if they feel you are working with them, not against them, towards their goals.

Finally, it is important to remember that discomfort and criticism of you as a therapist are not the same thing. Clients may be very uncomfortable but recognize that this is part of the process and not see this as a criticism of your ability. It can be helpful to check in with clients and facilitate feedback from them about how the therapy is proceeding. Many therapists are anxious about receiving feedback from a client and avoid asking because of the possibility of conflict or disagreement. Sometimes therapists respond defensively if a client does share feedback that seems critical. This can impair the client's learning about how to manage conflict and negotiate differences: we need to learn to recognize when our own vulnerabilities are triggered so that we can approach rather than avoid uncomfortable situations. For example, one supervisee was so unnerved by her client's telling her he thought she was "too passive" that she felt at a loss as to how to respond and was unwilling to discuss the client's experience in the moment. When she brought this situation to her supervision group, they were able to help her prepare to discuss this with the client at the next session.

Assessing what happens between sessions Assessing what happens between sessions can be structured or unstructured (or both) depending on whether or not you have assigned between session tasks. If you have assigned tasks, you might begin the session by asking about the client's experience and progress in relation to these tasks. If something of a higher priority is presented, you put that assessment aside while you focus on their current concerns. If the client has not performed the tasks, it is important to process with the client what kept them from engagement. For example, remember Juan from Chapter 8 who initially had difficulty following through with his assignment to increase outside activities between sessions? Do not expect continuous movement between sessions as change frequently really is a "three steps forward and two back" process. Rather than perceiving this as discouraging, focus on how you can help the client progress. Perhaps the client was not ready for these assigned tasks or does not yet have the skills or knowledge to complete them. Perhaps there are things going on in the client's life that you are not aware of that are affecting movement. Perhaps the tasks were not appropriate at this time for this particular client even though they may have worked with another client.

It is important to always keep in mind how the client can take what is being learned in therapy and apply it to his or her life outside of therapy. Talking about the time elapsed between sessions is therefore an important focus of any kind of therapy. This allows you to assess whether treatment is progressing towards agreed upon goals or whether the goals need to be modified or revised. The therapeutic relationship is sometimes the only supportive, empathic relationship the client has experienced. How does this experience affect his or her relationships outside of therapy?

Even when there are not assigned tasks, clients have experiences between sessions that are affected by the therapy. In these cases, discussing what has happened in the client's contexts and "internal life" (such as thoughts, feelings, story about self and others) is where the therapist finds the evidence of progress. As the client describes these experiences, you connect them to your conceptualization of the client's difficulties and contexts, seeking examples of change or of repeated problematic patterns that reflect your conceptualization.

EXERCISE 9.5 ■ Discuss in small groups your thoughts about how changes within the therapy sessions are brought out into the client's life. Do you think change happens primarily within the therapy room and is then generalized, or that the change process is guided within the therapy but happens primarily outside the therapy room? These thoughts will relate, of course, to your theoretical orientation and dimensional preferences. How do your thoughts and those of your peers differ?

Assessing what is happening over time Assessing over time is essentially assessing whether the treatment goals are being attained. Thus, this assessment is also inherently related to the conceptualization and the process of reconceptualizing.

John, age 52, would erupt by screaming and blaming his wife every time something displeased him at home, whether it was his wife, one of the children, or his dog. In couple's therapy, John learned to recognize his targeting of his wife and she, in turn, learned to respond rather than react so as to de-escalate battles before they got out of hand. While this pattern of couple interaction was never completely broken, the therapist was able to observe in sessions that the frequency and intensity of conflict decreased. The therapist also carefully evaluated with the couple whether the frequency and intensity of conflict had decreased between sessions and over time, and what new patterns were emerging. Even without formal measures, we assess progress through observations and clients' descriptions of their lives. We also assess progress through all of the other ways that we discussed gathering information (such as collateral sources, interactions between clients and therapists, formal assessment). Evaluation is just another type of information gathering; one that is more focused on evaluating change in the areas that are specified by the conceptualization and treatment goals.

As you are continuously assessing the process and outcomes of the therapy, the working alliance, and attainment of the therapy goals, you are thinking about how and when the treatment will end. There are two things that may result after achieving the initial goals of the treatment: 1) setting new goals or issues for focus in continuing therapy or 2) termination.

CHOOSING TO CONTINUE WITH NEW GOALS OR TO END THERAPY

The decision to begin a new phase of therapy with new (or modified) goals will be informed by the continuous gathering of information. Sometimes, clients find that they have met their initial goals, but that other issues have emerged during this process that they would like to work on in continued therapy. In addition to symptom removal, we need to think about how clients are developing cognitive and emotional self-awareness and sensitivity to others; how they are translating this into relationships outside of therapy; how they learn to make responsible life choices and problem solve; and how their functioning in all the contexts of their lives is improving. Are they satisfied? What do they think will get in their way as they move forward? What else do they want to have happen? How do they want to move on? Clients who have experienced positive changes in therapy frequently have found benefits in the change process offered by therapy that they want to apply to other issues in their lives.

At the same time, we need to be careful about developing dependency. Clients may fear that if they end therapy the changes will not continue or "stick." They may attribute the changes to the therapist or the therapeutic relationship, instead of to processes that they are responsible for and could maintain. We need to be aware of these possibilities, because in these cases clients may want to continue therapy not because they have new goals, but because they do not have confidence in the change created. Clients may also fear the loss of the therapy relationship. The therapy relationship is frequently one of intimacy (in a professional sense, of course) and trust; a kind of relationship characterized by acceptance and positive regard that is inherently rewarding. However, the rewarding aspects of the relationship should not mean a dependency.

One of the reasons for the push for empirically validated and evidence based treatments in recent decades is the criticism of never-ending therapy, a criticism particularly directed at psychodynamic and humanistic models. Because there is always the possibility of new insights, new areas to explore, and greater self-actualization, some therapies within these traditions would continue indefinitely. While some clients may desire this, the question remains whether this approach is best for the client's functioning, finances, and goals. How you answer this question relates to your idea of the overarching goal of therapy. For example, is one goal of therapy to improve functioning and experience enough so that therapy is no longer needed? Historically, only the affluent could afford the therapies that were then available and there did not seem to be any consideration for outcome assessment and fiscal responsibility. Trainees were taught how to develop relationships and treatment plans, but not how to gradually detach and end the therapy.

It can be the case, however, that choosing to continue therapy is not related to avoidance of termination, but is the best course for the client's health. Obviously there are contextual variables that will influence whether termination or continuation will occur, as well as the wishes of the client. Both the client and the therapist decide what is best at this juncture for the client and even if the therapist thinks the client could benefit by continuing therapy, the decision rests with the client. Even when the client desires to continue after initial goals are reached, it is frequently helpful to explore with the client not only the reasons for continuing therapy, but also the reasons why it may be helpful to stop therapy, at least for a time.

If the client opts for termination even though you as the therapist thinks he or she should continue, do not assume that you are not an "effective" therapist or that you did something "wrong." Focus on what you two have achieved, and expect that this work continues in many ways after treatment ends. Work with the client towards a planned termination that will help them consolidate the gains they have made and leave open the door for returning to therapy (with you or with another therapist) in the future, if they so desire.

TERMINATION

Termination typically means the ending of the therapeutic relationship and the treatment, at least for a time. Some therapists' theoretical orientation conceptualizes therapy as intermittent problem solving (Cummings, 2001); for them, the door is always open to return to do some more work, so termination is not a final "good-bye." Others have a theoretical orientation favoring on-going longer-term treatment as there is always more self-awareness and positive change to achieve. How your orientation meshes with your work setting and the client's circumstances will shape how you think about ending treatment. From some theoretical orientations, the therapeutic relationship never really ends; it continues symbolically within the client and/or may be picked up months or years later. You can liken this to the ending of a grade in elementary school where you are likely to have had a relationship with your teacher and classmates as well as developing a role in the classroom system. You carry some feelings with you, albeit negative or positive, but you move on to the next grade level experience. You may periodically drop in to re-connect with your former teacher if you had a strong connection.

The nature of termination will vary in relation to whether it is planned or unplanned and in which phase of treatment it occurs. Unplanned terminations occur when a client abruptly stops coming or for some other unexpected reason is unable to continue, or when the therapist becomes ill or for some other unexpected reason is no longer available. Pre-set planned terminations occur when a termination date is set ahead of time and is not dependent primarily on client progress. Examples of pre-set terminations are determined by agency policy regarding number of allowed sessions, limited third-party payor benefits, or the end of an academic training year. Co-planned terminations occur when both the therapist and client agree that it is time to end.

Terminations are frequently difficult for the therapist as well as the client. When we are evaluating whether to terminate or continue therapy, we need to be aware of our own responses and possible interdependency. If we do not know how to detach and end the therapy, we may avoid doing so for our own sake, rather than the client's. Unplanned terminations may challenge our basic sense of efficacy. But even planned terminations are challenging in some ways for most therapists, because a good therapist cares for his or her clients, and termination means the end (at least temporarily) of this intimate relationship.

EXERCISE 9.6 ■ Consider your experiences with ending relationships. These may include breaking up with an intimate partner, ending a friendship, moving from one teacher to another, and so on. See if you can identify what feelings and thoughts you had, what was difficult in the process, what was helpful in making you feel good about the relationship you did have, what you learned about yourself and the process of separation. Share these reflections with a partner or in small groups. How might your experiences with ending relationships affect your feelings about or approach to termination with a client?

Despite the difficult feelings that arise around termination, it is the therapist's responsibility to do his or her best to address these responses, to work with the client to evaluate and consolidate positive change, and to prepare the client to move on. Not doing so can undermine progress made and, for some clients, reenact harmful relational patterns.

Unplanned Termination

We try to avoid unplanned terminations by checking with the client regularly about how "things are going." But, sometimes, we are taken by surprise. One supervisee was working with a young woman who was suffering from post-traumatic stress syndrome due to childhood incest and previous therapist sexual abuse. After eight sessions, while working on anger issues, the therapist suggested the client write a letter to the previous therapist about her feelings of betrayal—not to be sent, but as a way of expressing some of the feelings that were difficult for her to verbalize. The client did not return to therapy and refused the therapist's invitation to talk about her experiences in this current therapy for no fee. The therapist was quite distraught, wondering if her timing of the intervention was off and how she had misread the client's ongoing reports that she felt safe and was pleased with what was happening in this treatment. Unplanned endings leave unfinished business for both the therapist and the client, and sometimes there is no way to achieve closure. We recommend in such cases that the therapist write a letter to the client offering to meet without charge to discuss the client's progress and the client's concerns about the therapy and offering to help with a referral. This kind of letter communicates caring and concern, an interest and willingness to learn how and what was not working for the client, and a willingness to take responsibility for anything that the therapist may have done that was harmful to the client.

We need to do our best to avoid unplanned terminations by carefully conceptualizing our clients, continually evaluating our conceptualizations and the effects of our interventions, and working to build a strong working alliance with our clients. Strategies such as preparing clients for therapy, providing case management to clients during difficult times that might challenge continuing therapy, providing appointment reminders, and explicitly engaging clients' motivation (for example, through motivational interviewing) can be helpful in avoiding premature termination (Ogrodniczuk, Joyce, & Piper, 2005). However, the therapist can never carry the primary responsibility for change. There are inevitably instances where unplanned terminations occur because the client has decided that he or she is not ready or willing to invest the time, energy, motivation, money, or other resources into changing their current situation.

We have had many supervisees who have had clients who come to therapy for one or two sessions and never come back. These supervisees frequently worry that they are doing something "wrong" and clients are dropping out because they are "bad" therapists. But when we ask about dropout rates in the organizational settings in which they work, it becomes clear that even the most experienced and expert therapists in these settings have similar dropout rates, which suggests that it is more about the clients and client circumstances or the structure than about therapist skill.

Furthermore, premature unplanned terminations are sometimes related to issues we cannot control, such as transfers that have been poorly conducted. One of us had a case, for example, where the prior therapist did not terminate well with a client diagnosed with Borderline Personality Disorder. The previous therapist was having sporadic contact with this client through letters and telephone calls that the client described as very positive. The client repeatedly stated that the former therapist was the best and complained from the second session in vague ways that the current therapist was not helpful, although she refused to discuss any other aspects of the prior therapy or therapist with the current therapist or to offer any details about what would be helpful in the current therapy, in spite of continued invitations to do so. Almost from the first session, the client came very late to sessions, canceled multiple sessions at the last minute, and did not respond to the therapist's calls. The therapist continued outreach to the client, attempted to supportively explore the reasons offered for last minute cancellations and lateness (for example, the client's work demands), and invited the client to share other aspects of her life and concerns. After several weeks characterized by this behavior, the therapist asked a tentative question of whether the client felt that the lateness and cancellations might reflect some concern that the therapist was not as helpful as the client would like. The client responded by yelling at the therapist that she did not understand anything and was a terrible therapist, and then storming out.

EXERCISE 9.7 ▪ Consider the situation with the therapist in the previous paragraph. What are your thoughts and feelings about it? If you were supervising this therapist, how might you respond to the supervisee if he or she shared this incident with you?

Unplanned terminations may also occur in less abrupt fashions such as when a client needs to unexpectedly relocate, when the client becomes ill, when the therapist becomes ill, and so forth. If the termination is completely abrupt with no warning, then letters such as those described above are important follow-ups. If the unplanned termination occurs because of a crisis in the therapist's life, it is an ethical imperative to ensure that clients receive appropriate referrals for continuing care, and that all efforts are made to terminate in a fashion that is helpful to the client. Telephone sessions to terminate well may also be an option. In other cases, there may be a bit of warning (such as a few weeks), in which case the therapy will need to move into a termination phase (see Planned Termination, next section) while also focusing on transferring the client and/or arranging other services to meet the unmet goals of the therapy.

Unplanned terminations in the beginning phases of therapy that are not related to ruptures in the working alliance are frequently not as complex as in later stages because the relationship is not as well developed and there may be little, if any, real progress upon which to reflect. In later phases, there is typically a stronger working alliance and commitment to the therapeutic process by both therapist and client. Thus, the feeling and processing may be more intense for both parties.

Finances and Termination

It is not unusual for clients to require more sessions than allotted by agency policy or third-party payors. In most cases, the number of sessions allotted should be known ahead of time and the therapy planned accordingly. In private practice and in many organizations, maintaining billable hours is a salient concern. So it behooves us to consider session limitations in setting goals and to not attempt to "seduce" clients into remaining in treatment for any reasons other than what is in the client's best interests. This, of course, is also an ethical imperative. When clients need more than the number of sessions and the conceptualization and treatment planning make this clear, the clinician advocates for his/her clients as much as possible with the third-party payors and/or with the treatment site. This takes time—non-billable time—but it is part of responsible, ethical treatment. Treatment planning and goal setting may need to be planned in a flexible manner, as it may not be clear how many sessions will eventually be approved. Careful prioritization of goals, as well as shorter-term goals that build upon each other over time, are most helpful in these circumstances.

Increasingly with the shrinking economy, we find that clients are losing their jobs and some of their benefits. Or we may believe that additional sessions will be approved only to have this denied. These circumstances may lead to an unplanned termination because of financial reasons. If all else fails and we are seeing a client who needs additional sessions that cannot be paid for, we may need to provide pro bono (without charge) services until we can make an appropriate referral or find appropriate services for the client. Remember, codes of ethics prohibit "abandonment" of a client by a therapist (Vasquez, Bingham, & Barnett, 2008). We need to creatively find ways to provide services. Some clinicians may transfer individual clients to groups with lower costs, shorten the length of a session to expand the number of sessions, or reduce fees for a period of time. Each set of circumstances is different. The kinds of questions you might ask yourself include:

1. What is in the best interests of the client at this point in time that I am able to do?
2. What are the client's options?
3. What options can I utilize? How can I help the client and myself to make the best of an unfortunate situation?
4. How can I help the client adapt to changing circumstances—for example, focusing on job-seeking skills, changing vocations, obtaining new training, and so on?
5. How can I emphasize the client's strengths and what the client has gained to date in treatment to navigate changing circumstances?

Planned Termination

Planned termination may occur because the therapist and client agree on a certain number of sessions, and/or agree that the goals have been met.

Planned termination may also occur because third-party payors and agency calendars have a set number of sessions or because of termination dates such as the end of an academic or training year. In theses case, termination is planned but also premature. Planned termination is typically structured in that both the therapist and client are mindful of the length of treatment and planning for relapse prevention or follow-up activities throughout the treatment. Therefore, termination does not occur abruptly or come as a surprise.

The process of termination. What is most vital about termination is the process or manner in which it is discussed. Optimally, termination is a gradual process. Depending on the orientation of the therapy, the client is usually given the opportunity to share his or her feelings about ending the therapeutic relationship, as the loss of this relationship may bring up feelings related to loss or grieving of other relationships. The therapist may also share his or her thoughts and feelings about the separation or ending. At least a couple of sessions should be used to review what brought the client into treatment, what he or she has learned in therapy, how he or she has grown, and how he or she will maintain the changes from therapy. The therapist should not bring up major new issues associated with intense emotion during this time. Optimally, termination can be framed as a new beginning for the client, a type of graduation related to achieving goals.

For both therapist and client, other experiences of attachment and separation may shape these thoughts and feelings and the termination process may be more or less intense, depending upon the people involved, the outcomes of the treatment, and the circumstances. While these related experiences might be evoked from the client and discussed in session, the therapist should discuss his or her responses in supervision or consultation. It can also be helpful to discuss in supervision the ways that this therapy and this client have contributed to the growth of the therapist. Some aspects of therapist process that are not countertransferential may be brought back to sessions with the client, depending on the theoretical orientation, conceptualization, and needs of the client. It can be helpful to let clients know that we will continue to value them and that we appreciate what we have gained from working with them (Penn, 1990).

Planned termination due to completed goals. How do we know when a client is ready for termination? Is it client- or therapist-initiated? Either can occur. Sometimes clients express a belief that they have resolved whatever problem(s) brought them into therapy and they want to stop or at least take a break. Typically, the therapist respects the client's wishes unless there is reason to believe the client is evading some issues, which the therapist can then raise as a therapeutic concern. Sometimes, the therapist initiates a discussion and planning for termination because he or she thinks that the client's goals have been achieved and it is time for the client to learn to function without the therapist's support. Usually there are clear cues from the client when the treatment has been successful and the client has internalized the external support from the therapist.

For example, the client reports "feeling better" and describes changed relationships, behavior, and patterns of thinking or feeling that are generalized to multiple contexts; the therapist observes positive growth; collaterals report improvement in school, in health, and in family. Many clients report that they "hear the therapist's voice in their head" when they are feeling stressed or engaged in problem solving. By this, they mean that they have internalized the positive voice, perspective, or problem solving approach of the therapist to serve as a continuous role model. In these cases, termination is indicated because of successfully completed goals, and the client and therapist should set a time to end and begin the termination process.

Planned premature termination due to the end of the training year. Therapists in training frequently conduct planned premature terminations due to the end of the training year. Although planned, this kind of termination is premature in that the treatment goals are not fully achieved. However, the ending should not come as a surprise to the client, who should have been advised of the time-limited nature of the relationship from the beginning. Trainees typically begin termination discussions four to six weeks prior to the end of the training period, talking with clients about transfer, ending therapy, or other options.

Trainees are frequently challenged by termination not only because of the inherent challenges discussed above, but also because this kind of planned premature termination is a moment when the needs of the therapist (that is, training schedules and structures) intrude and "trump" the needs of the client (that is, to continue). In addition, this kind of termination can be particularly challenging for the training therapist because it is not only one termination with one client, but multiple terminations with multiple clients, as well as terminating with supports such as supervisors, all occurring simultaneously. In spite of these differences, many of the tasks of premature planned termination are in many ways similar to a planned termination due to a completed therapy. (See Penn, 1990 for an overview of issues and approaches and Gelman, Fernandez, Hausman, Miller, & Wieiner, 2007 for a consideration of the role of supervision in trainees' termination experiences.)

The multiple and simultaneous nature of multiple terminations, accompanied by the fact that termination of a therapy relationship is a relatively new experience for trainees, may contribute to a greater likelihood of countertransferential responses. It is important that trainees anticipate this and carefully attend to the ways in which their own feelings about endings and loss (as described above) are affecting this last stage of therapy with their clients. Another difference in the process is that the therapist's trainee status can be highlighted by the need for termination. Our experience is that this rarely takes the form of the client questioning the skill or expertise of the training therapist, but much more frequently takes the form of the client being curious about the next phase in the therapist's professional career or development; this is usually related to the real relationship, as clients have come to care for the therapist as people. Theoretical orientation and the policy of the organization or supervisor will affect how these issues may be dealt with, but in our experience, it is usually most helpful to simply describe in general terms what the next step will be (for example, to start a postdoctoral fellowship in another organization, to begin a research position, and so forth). Another major difference is that clients who are seeing trainees and experiencing a planned, premature termination are frequently transferred to another therapist.

Transfers. In many sites, termination occurs at the end of an academic or training year, and it is expected that the client will continue with a new therapist at the start of the new academic or training year. Sometimes it is best for the client for the departing therapist to arrange for the client's transfer to a therapist with particular characteristics rather than random assignment. In some situations, that is possible and in some it is not. But it is important for the terminating therapist to leave written records and to advocate strongly for a positive transfer, that is, a therapist that is particularly suited to the client. The subsequent therapeutic experience is likely to be very different, and the goals and nature of the treatment will be determined within the context of the new therapeutic relationship. As emphasized throughout this text, therapy is a continuous, circular process and different experiences can be equally valid.

Referring to the earlier example of an ineffective termination prior to transfer, let's talk about how we can prepare clients for transfer. Some clients have been "in the system" for a long while and are accustomed to working with a new trainee each year, while others may have never experienced a transfer to a new therapist. In either case, it is helpful to review with the client which goals have been attained, which remain to be attained, and what new goals might now be appropriate. It is also important to prepare a report with this information for the next therapist. It is also important to go over with the client what has worked and what has not worked in the working alliance and to prepare the client for a different experience with another therapist. In other words, we do not want the client to think that the therapy process will be *the same* with someone else; we want to highlight the value of different experiences and that there are multiple alternative routes to the eventual destination.

Trainees often experience dismay because they do not know who the new trainees will be and, therefore, they cannot aid in selecting a desirable fit. And, in some cases, there will be a lag between the time of termination with a trainee and a new assignment. In the latter case trainees may arrange for interim programs, such as groups or camps or some other community resources. If these arrangements cannot be made, it can be frustrating for the trainee who understandably feels that the progress achieved might not be maintained. There may be creative strategies that the therapist could suggest to the client for this interim period—for example, keeping a journal to be shared with the new therapist, selecting a friend or family member to talk to, or other specific activities. Some clinicians co-author a personalized manual with the client during termination, with suggestions to guide them through anticipated difficulties. For the therapists themselves, it is important to use supervision to discuss feelings about endings and limitations on best practice follow-up, as well as feelings about their clients being seen by new and different therapists.

Termination with Nancy

After Nancy's eating disorder treatment and upon the return of the therapist, she and her therapist began to focus on post-college planning. While this was briefly mentioned during Nancy's final semester, focus had been on completing the semester in order to graduate and reducing the drinking and binging. The eating disorder treatment program had been immensely helpful; Nancy had actively participated, enjoyed the structure of the group, and even made a friend among the other patients. She seemed more relaxed and in control than at any time earlier.

The therapy had a primary goal at this point: to help Nancy leave home and live independently. Originally, Nancy had thought that she would go straight on to graduate school, but her parents were not encouraging, wanting her to get a job and to live away from home. This was difficult for Nancy as her feelings of abandonment and rejection from her family intensified. With her therapist, she accepted this as a goal and was able to consider her options. With a great deal of diffidence, she researched possibilities and applied for and received a paid internship in another city. Once this became a reality, she became panicked about how and where she could live. She told her therapist that she wished her mother would help her to plan a budget and give her some tips about moving out, but her mother refused to engage with her, replying that she herself had done it by going to boarding school at the age of 10 and she expected Nancy to do it alone. Again, without any family support, she and her therapist considered many possibilities, finally deciding upon temporary living arrangements at a YWCA in the new city. Once these arrangements were in place, Nancy and her therapist began to talk about termination.

With a date for termination set by these plans, the therapist reviewed Nancy's original goals as well as her own. While Nancy was not "happy," she certainly expressed feeling stronger and better about herself except for periodic anxiety attacks. She had achieved her academic goal of graduating and she had come to realize that her goal of persuading her parents to give her more financial support was not going to happen. However, rather than be angry about this, she had begun to realize that she could not expect that and she needed to learn to manage her money and live within her income. She even suggested that she use some of her graduation money to move, which she had refused to even consider months before. Nancy was also beginning to utilize some of the strategies suggested by her therapist to manage her anxiety, which was becoming less intense.

The therapist felt that Nancy was regulating her emotions better, had developed awareness of her feelings and beliefs, and was making concrete efforts to change her relationship behaviors. The therapist felt that Nancy needed support while launching, but she felt that the highest priority was helping Nancy leave what really was a discouraging home situation. She was impressed by Nancy's active initiation and participation in the planning process and by her newfound ability to identify her anxieties and feelings and to begin to incorporate into her life some of the stress management strategies they had been working on. They talked about their mutual sadness about the separation, and the therapist offered periodic telephone sessions after Nancy moved and help in locating a new therapist where Nancy would be living. Nancy was not sure she wanted to see another therapist, but expressed appreciation for her therapist's concern.

At the last two sessions, they reviewed Nancy's new learning: 1) she was able to recognize and acknowledge when she was behaving in ways that were alienating to others or self-destructive to her own goals; 2) she was determined to move out on her own without asking anything of her parents (in the past when she requested help, she was refused and shamed for asking); 3) she was more aware of her role in conflict with others; 4) her entitlement thinking had changed dramatically as evidenced by what she expressed and did; 5) she was taking better care of herself in terms of eating and drinking. What was most difficult for Nancy was coping with the pain she experienced by her parents' cold lack of support. She was afraid that once she left home, they would forget about her and not make any efforts to include her in the family. The therapist was not sure this would not be the case, but she was able to help Nancy accept that she needed to get on with her life, and that she could hope and wish for more accepting family, but could not make them be who they were not. Nancy and her therapist decided that Nancy would initiate contact with her family once a week, but also discussed reasonable expectations for her family's response.

There were some goals that were not fully met. One was that while Nancy was able to begin to consider others' perspectives and loosen some of her entitled cognitive schema, she tended to revert back to her earlier ways of thinking and behaving when she became stressed. She did have tools to manage anxiety, but did not always recognize her feelings early enough to utilize these tools. Certainly, her emotional regulation had improved and she had more awareness of her emotions. A major goal that was not realized was changing relationship patterns. She still suffered social anxiety, felt distrust of others, and was very vulnerable to rejection. She was aware that at the YMCA she would be in a structured residence with opportunities for group and social activities and promised herself (and the therapist) that she would make efforts.

The final session was teary. Both the therapist and Nancy were sad and able to talk about the sadness and how meaningful the work they had done together had been. But Nancy was resolved, thanked the therapist, and said she wanted to keep the door open for telephone or email contact should the need arise. She promised to come back to see the therapist when she returned home for a visit. The therapist had some anxiety about how Nancy would fare, but felt that this was an important step and that she and Nancy had done some effective relapse prevention work to prepare her.

FOLLOW-UP AND RETURNING CLIENTS

Follow-up

While trainees may not have the opportunity for follow-up, many therapists suggest a check-in several months after termination to check on continued progress. This can be done in person, over the telephone, by completion of follow-up forms, or by another therapist at the organization. Typically, who will do a follow up depends on the policies of your site and on the particular circumstances of the therapeutic relationship. Many therapists find that follow-up is beneficial to assessing outcomes of treatment over time and to demonstrate interest and support. Obviously, the purpose of follow-up is not to encourage a client to return to treatment.

Nancy indeed did schedule two telephone sessions the first month she was gone and then did not contact her therapist until she came home for Thanksgiving. She stated that she "loved" her internship and was hoping to be put on staff after her six months were up. She was not very happy at the YWCA, but was hoping that a friend from high school would get a job in the same city after the first of the year and that they could find an apartment together. She did report that her parents had invited her home for Thanksgiving and, while they did not initiate contact with her, they were friendly and cordial when she called home. Her therapist felt reassured by these follow-up sessions and pleased that Nancy was indeed able to live independently and that her symptomatology was greatly reduced.

Returning Clients

It is not unusual if you are a therapist practicing for a long time in the same community for clients to return awhile, often many years. Sometimes, this is just for a one-time problem solving session or a sort of check-up; sometimes, new circumstances have motivated them to do another piece of work. As their previous therapist, you have the advantage of an earlier therapeutic relationship and treatment experience upon which to build, which you do not have with a new client. But while the relationship has already been formed, you begin with a returning client as if they were a new client with regard to the process of conceptualization and treatment.

Sandra, age 34, heard through a friend that the therapist trainee she had seen four years ago at a community mental health agency had completed her training and was now a member of the agency staff. She requested an appointment as she was dealing with the disability of her one-year-old son and was re-experiencing the panic symptoms she had presented with during her initial therapy. She had seen a therapist at her HMO, but did not find the sessions to be helpful. The therapist was looking forward to this meeting as she felt they had worked well together in the past and she was eager to learn how Sandra's life was progressing. At the time they were originally working together, Sandra was just beginning to date the man she eventually married. At this meeting, Sandra spent half of the time updating the therapist about the intervening four years and expressed relief at being "back" even though the physical office was different. The working alliance had already been established, but the current context had to be thoroughly explored in order to develop new goals.

FINAL THOUGHTS

Therapy is a process of growth and learning to be a therapist is also a process of growth. As you "terminate" this particular learning experience, you may experience mixed feelings and thoughts about your learning. There are likely things you have learned and ways that you have grown that you feel are good achievements and progress toward your goal to become a good therapist. There are also likely things that you are still uncertain or apprehensive about, and things you have identified as areas for future growth.

EXERCISE 9.8 ▪ How do you feel about ending this book, this class, this moment in your training as a therapist? Many of you will likely be moving on to practice experiences where you will have less guidance and/or supervision and may not be simultaneously reading books such as this one about conceptualization, treatment, or developing as a therapist. How does that prospect feel to you? If you are taking a class, how does it feel to be moving on from your teacher and these particular peers in this setting?

EXERCISE 9.9 ▪ Write down five things you have learned about yourself as a therapist. How have your thoughts and feelings about the process of therapy and yourself as a therapist developed? What are five things you want to continue to work on in your development? Compare your lists with those of one or more others and then discuss possible strategies for consolidating your growth and continuing this learning.

We shaped the last two exercises to parallel the process of termination, as described in this chapter. Obviously, there are no final and absolute conclusions we can provide you with, just as there is no final word that you can offer your clients. The process of growth is a continual one.

For us, writing this book has also been a process of growth. We have learned from each other and from the process of articulating our beliefs and experiences, and linking these to research, theory, and clinical findings. We have enjoyed the process of imagining you (the readers), thinking about the past and present trainees with whom we have worked, and hearing from our current trainees how these chapters have been helpful to them (and how to make them more helpful to you as we engaged in revisions). We believe that we, too, are now better therapists.

We hope that we have accomplished our goals: a) to help you to view psychotherapy as a complex interpersonal process dealing with the multilayered experiences characterizing human behavior, development, and change; b) to develop an understanding of why it is important to develop an integrative theoretical orientation and a continually changing conceptualization to guide treatment planning; *and* c) how to go about doing this. We do believe that, regardless of their training socialization, all therapists need to be flexible and open to multiple realities and possibilities; we need to make continuous critical appraisals of our always developing personal and professional values, attitudes, thinking, and feeling; we need to be intentional about our work; and, finally, we need to be aware of the connections between our personal and professional values, beliefs, and ethical positions; between theory and practice, between self-awareness, empathic attunement to clients and continual conceptualization and treatment planning.

Like you, we look forward to continuing growth. Our own ideas about human behavior, health, pathology, and treatment from an ecological perspective continuously evolve as science and practice advance in an increasingly complex global context. It is intriguing to think about the possibilities of new findings and what that will mean to us as clinicians, as teachers, and as supervisors. Our training, personal, and professional development and activities are intertwined and ongoing. This allows for intellectual and emotional stimulation and challenge as we improve our effectiveness as helpers. We thank you for sharing this part of the journey with us.

REFERENCES

Cummings, N.A. (2001). Interruption, not termination: The model from focused, intermittent psychotherapy throughout the life cycle. *Journal of Psychotherapy in Independent Practice* 2(3), pp. 3–17.

Gelman, C., Fernandez, P., Hausman, N., Miller, S., & Weiner, M. (2007). Challenging endings: First year MSW interns' experiences with forced termination and discussion points for supervisory guidance. *Clinical Social Work Journal* 35(2), pp. 79–90.

Ogrodniczuk, J.S., Joyce, A.S., & Piper, W.E. (2005). Strategies for reducing patient-initiated premature termination of psychotherapy. *Harvard Review of Psychiatry* 13(2), pp. 57–70.

Penn, L.S. (1990). When the therapist must leave: Forced termination of psychodynamic therapy. *Professional Psychology: Research and Practice*, 21(5), pp. 379–384.

Vasquez, M.J.T., Bingham, R.P., & Barnett, Jeffrey E. (2008). Psychotherapy termination: Clinical and ethical responsibilities. *Journal of Clinical Psychology* 64, pp. 653–665.

Appendix A

A Brief Review and Application of Established Theories

Theories of human behavior and psychotherapy reflect the zeitgeist of the era and region in which their originator and later developers lived. They reflect these proponents' personal temperaments, characters, values, attitudes, and beliefs within the socio-political, economic, and historical contexts in which they live or lived. To consider theory apart from the context of the theorist in his or her contexts gives us a distorted view not only of the theory, but also of its applicability in today's time and cultures. This also means that you must consider your theory or orientation in relation to the contexts in which you are living.

Psychologists construct theories that serve as explanations for their experiences as well as their observations of others' experiences. Since their experiences and observations have been filtered through their own lenses, their theories cannot be separated from their personal values and beliefs, which, as mentioned, have been shaped by sociocultural context. What is a "psychological theory" if not a hypothesis or speculation about people's behavior and ability to grow and develop? Theories are not facts; they are always subject to modification and revision. Scientific theories, which many psychotherapy theories claim to be, are based on an initial set of assumptions and observations that are subject to empirical testing in order to be further refined and defined. The hope is that the empirical testing will result in what (Kuhn, 1962) terms a "paradigm," a dominant systematic position. To date, we do not have one fully acceptable paradigm to explain human behavior and psychotherapy. What we have are process models (cognitive maps) based on philosophical assumptions with some elements of scientific theory. These models guide the process of psychotherapy and can be evaluated only on the basis of treatment success. Furthermore, their success will vary with different clients, at different times, in different contexts.

Each of the major models of psychotherapy has devotees who believe that their view is the only correct view. This type of doctrinairism has probably done more harm to the development and credibility of the psychotherapy field than any other single variable, because it has reinforced turf competition and dichotomous thinking such as right or wrong, science or art, good or bad. Not all of the traditional theoretical orientations which claim to be "complete packages" address explicitly how people develop, how psychological problems develop, how and why these psychological problems are maintained in relation to different contexts and, therefore, what is the best way to intervene for positive psychological change.

In the sections below, we briefly review some of the major theoretical approaches and apply these to understanding the case of Nancy. We are assuming that you have some grounding in these theories and the critiques that have been made of them. In such a brief review, it is inevitable that each approach will be vastly oversimplified. It is important to remember that within each major approach, there is considerable variability, and that there are areas of overlap between approaches, particularly for any given therapist. For example, some cognitive-behavioral therapists emphasize behavior more than cognition. Furthermore, most cognitive-behavioral therapists strongly attend to a client's affect, but they see this affect as related to thoughts and behaviors, and so their interventions are usually not explicitly aimed at addressing affect directly. A major thesis of this book is that theoretical orientation, for most therapists, is very complex, nuanced, and frequently influenced by multiple approaches, even if a single approach is claimed as the primary theoretical orientation. We encourage you to recognize that these brief reviews are not at all representative of the ways in which therapists actually utilize these approaches for case conceptualization and treatment planning.

Although we have tried to avoid using comparison to define approaches (as this tends to further dichotomize and oversimplify), in such a brief review it is also inevitable that the differences between approaches are emphasized, as being so brief means that the unique rather than shared characteristics receive more emphasis. Also, although there are many critiques and criticisms of each approach, both from other approaches and also in relation to how well each approach addresses contextual variables and complexity, our goal here is not to critique, but to briefly review major approaches. Our purpose is to remind you of their foundational beliefs in order to encourage you to consider how these established approaches will contribute to your own theoretical orientation.

REVIEW OF THEORETICAL APPROACHES

Psychodynamic Theories

All psychodynamic theories believe in unconscious motives outside normal awareness. They focus on infant and child psychological development; one must successfully navigate a stage before progressing onto the next stage.

Psychodynamic theories, in general, focus on the recognition and interpretation of unconscious id libidinal and aggressive drives and the ego defenses (primarily projection and repression) against them as well and/or on relational drives and the connections between childhood relationships with primary caregivers and current relationships with significant others.

Classical psychoanalysis, an individual drive model, posited that psychopathology derived from conflict between the psychological structures of the id, ego, and superego and that neurosis derived from conflict between these structures catalyzed by the oedipal conflict. The impulses of the id and the reactive punishment of the superego need to be brought under the control of the ego, and the therapist's interpretations aim to increase patients' ego controls. Freud's (1936; 1943; 1953; 1964) work was based on two fundamental hypotheses: 1) psychic determinism, which means that nothing happens randomly or by chance; each psychic event is determined by preceding ones so the therapist always asks "What caused it?" "Why did it happen so?" and 2) the existence of unconscious mental processes (validated recently by brain imaging technology; Gulyas, 2009; Shevrin et al., 1996) which have powerful influences on conscious thoughts or behavior in both healthy and unhealthy populations.

One of the main offshoots of psychoanalytic theory is the individual psychology of Alfred Adler (1925). Adler's basic concepts include: 1) people are not a collection of instincts and drives but rather a whole open system; 2) we can only study people by their actions and relationships within social groups; 3) individuals are responsible for their own development; 4) we cannot understand people without attending to their life style patterns, which reflect their feelings; 5) an individual's character is determined by societal, love-related, and work variables; 6) an individual's defense mechanisms include compensation, resignation, and over-compensation, that is striving for power and superiority to compensate for feelings of inferiority. What Adler contributed to psychoanalytic theory was an optimistic, social focus on personality; people are viewed as social beings needing a sense of belonging. This switches the focus from sex and aggression to social factors.

More current psychodynamic theoretical developments emphasize how attachments and relationships (rather than drives and structural conflicts) are the basis for human behavior; object relations was first developed in Great Britain during the late 1930s and self psychology first developed in the United States in the 1940s (Okun, 1990). Thus, as opposed to Freud's intrapsychic model, object relations is an intra- and interpersonal theory where psychopathology derives from developmental arrest due to unsatisfactory attachment/separation experiences in infancy and early childhood. Neurosis, therefore, is derived from pre-oedipal conflict related to relational interactions, earlier than Freud's oedipal conflict. Object relations theory focuses on peoples' internalization of real relationship experiences. People form internal images of their relationships with other people through their own filters. Thus, different individuals may internalize the same relational interaction differently. Kohut's (1971; 1977) theory of self psychology helps us to understand an individual's need for empathic connections with primary caregivers; a cohesive, well-integrated self can develop if the child has been appreciated, idealized, and mirrored by loving, supportive parents.

Like all psychodynamic models, object relations and self psychology are developmental models. In both object relations and self psychology, how one creates internal images of self, others, and the world from early parent–child interactions determines later intimate relationships The therapeutic outcome depends on an empathic connection between the therapist and patient to repair developmentally difficult object relations and self-concepts. Development of interdependent mutually gratifying relationships with self, with others, and with the world is the major focus and goal. In this model, aggression is not innate; rather it is a reaction to frustration from an internalized bad object.

EXERCISE A.1 ▪ We all have internal "scripts" (fantasies) about who we are, how others are, and how our life is. See if you can get in touch with your internal script—the role you have for yourself and for significant others in your life. How does your internal script jibe with your objective experience? What would you like to change and how might you do so? The purpose of this self-reflection is for you to become aware of some of the complexities in your internal life in relation to your external life.

EXERCISE A.2 ▪ Psychodynamic theories share an assumption that there are processes and relational influences that are unconscious that can be brought to consciousness. Have you had experience with becoming aware that your behavior was influenced by a desire or an idea about relationships that you only became aware of later? Have you had an experience of realizing that your responses to someone (for example, your partner, a close friend) were being shaped more by your idea about how "people" act based on your internalization of relationships rather than your actual interactions with that specific person? Discuss this in small groups.

Psychodynamic Conceptualization of Nancy A psychodynamically oriented therapist would focus on developing a working alliance with Nancy in order to help her re-experience the critical events of her childhood. The assumption would be that Nancy's current difficulties were caused by the attachment and separation traumas she had experienced with her parents during early childhood. Nancy had not received the empathic attunement that would allow her to develop her "true" self. Thus, Nancy currently is seeking attachment with a boyfriend to provide object constancy and to help her to feel valued and cared for. Nancy's inner psychic reality—consisting of a fragmented, damaged, weak self formed in early childhood—has become her external reality where she avoids interpersonal contacts because she fears rejection.

A psychodynamic therapist would work with Nancy to help her gradually understand her motives and her human relationships. This would be achieved through an intense, trusting relationship wherein Nancy can experience a positive attachment to the therapist and gradually internalize a more positive view of self, others, relationships, and the world. They would explore Nancy's early attachments and separations, early sexual and aggressive traumas (real and fantasized)—in other words, how Nancy's pre-oedipal and oedipal conflicts were resolved or not resolved—an attempt to make Nancy more consciously aware of these issues and how they affect her current experiences. The psychodynamically oriented therapist would most likely focus on the transference relationship, whereby Nancy re-creates with the therapist critical elements of her past significant relationships, particularly her yearnings for closeness and approval and her fears of rejection in order to become aware of the ways in which she enacts her internalized assumptions about relationships. The desired outcome would be for Nancy to achieve intellectual and emotional insight so that she could be free of her pain, make more conscious and considered decisions about her behavior and relationships, and have more energy to engage with others and in work with mutual gratification.

Cognitive-Behavioral Theories

The cognitive-behavioral model is the rapprochement of traditional behavioral theory and cognitive theory. Both theories developed in the United States in direct repudiation of the influence of the Western European psychoanalytical model. Cognitive-behavioral models differ from the psychodynamic models in the following ways: 1) they seek to be objective and empirical in contrast to the more subjective psychodynamic approach; 2) they are based on learning theories and environmental influence in contrast to the medical basis and relative intrapsychic emphasis of psychodynamic approaches; 3) they are symptom-focused in relation to intervention planning; and 4) they emphasize concrete rather than abstract, cognitive or behavioral rather than affective, and, in some cases, rational rather than irrational aspects of experience.

Behavioral theories posit that behavior is under stimulus control, not biologically or psychically determined. People behave in certain ways because they have learned that those ways are rewarding and developed patterns of thought that support these behavioral responses. However, while certain behaviors may be rewarding in some contexts or times, they may not be the best choice when circumstances change. Furthermore, the behaviors or assumptions may be based on inaccurate connections between events during learning. Cognitive theories posit that learned experiences lead to ways of thinking that then affect behavior, rather than behavior being directly shaped by environmental contingencies. Beck's (1967; 1976) cognitive theory, for example, suggests that individuals can choose, from a vast number of options, the option that appears to be in their own best interest. Thus, self-defeating behaviors come from an inability to conceive of, act upon, or carry out more constructive alternatives. For Beck, the world is a series of positive, neutral, and negative events. Individuals interpret these events with a series of thoughts, which lead to behaviors and affect. Individuals have the capacity to modify their thoughts and, thus, to modify the ways in which they act on the events in their environment.

The unique contributions of approaches based on learning theory are their focus on overt response, their reliance on conditioning, and their notion of reinforcement. The recognition that behavior is learned and that cognitive processes regulate the impact of environmental forces on behavior is a major contribution of this model. Human beings are active agents of change and, therefore, have the capacity for self-motivated behavior change. One perceives, interprets, and transfers environmental stimuli in relation to existing cognitive structures. Then, one behaves in response to these interpretations and transformations and in response to environmental contingencies (rewards and punishment). Thus, psychological disturbance is caused by one's attitudes and perceptions and by one's attribution of meaning to experiences.

The belief that all behavior is learned and can thus be unlearned or modified offers hope to those who feel overwhelmed by behavioral problems, just as the cognitivist's belief that one can be in control of one's thoughts, rather than controlled by them, offers hope to those stuck because of dysfunctional thinking. The cognitive-behavioral approach is deterministic in that the individual is still partly subject to the impact of the environment. However, its notion of the individual as an active contributor rather than a passive reactor is more optimistic than a more completely deterministic notion of human nature. In cognitive-behavioral therapy, the focus is on developing skills through techniques such as modeling, psychoeducation, skill training, and practice. The therapist has the primary responsibility for accurate appraisal of the problem, selection and teaching of skills, and outcome evaluation.

EXERCISE A.3 ▪ In partners, take turns with the following: 1) imagine the last time you were upset; visualize the scene in as much detail as possible; 2) describe the situation and details to your partner; 3) how did you experience this upset, that is with what part of your body?; 4) elaborate your feelings, thoughts, and behaviors. Now, with your partner see if you can identify the thoughts that underlay your upset. How could you restructure those thoughts?

EXERCISE A.4 ▪ Think about a behavior that you would like to decrease or increase (eating, daydreaming, exercising, procrastinating, smoking, and so on). What made you start doing that behavior? What are the rewards in continuing to or not continuing? What are the reasons to increase or decrease the behavior? In a given moment of decision (to smoke or not to smoke, for example), are you aware of making a decision? What is (or would be) your thinking about the decision if you are/were aware?

Cognitive-Behavioral Conceptualization of Nancy The cognitive-behavioral therapist would work to develop a collaborative relationship with Nancy so that both of them could determine precise goals for change. The therapist would utilize psychoeducational strategies and assignments, particularly in the area of relationships and would structure particular behavioral learning experiences. The cognitive-behavioral therapist would focus on Nancy's ingrained belief systems and engage her in cognitive restructuring exercises. Nancy would be given homework exercises to identify her dysfunctional thinking and to learn to refute and challenge these thoughts. Some of the more obvious dysfunctional or irrational thoughts Nancy has are: 1) things should be the way I want them to be; 2) people (my family) should do what I want them to; 3) I should have what I want when I want it.

While the past might be explored in order to trace the path of her dysfunctional thinking and its associated emotional pain, the focus would be on the effects of these cognitive distortions on Nancy's everyday life today, not on the etiology of these issues. There would be no seeking of early-childhood causes of her current feelings and no exploration of her early family relationships. Nancy's current family relationships, her relationships with her suitemates and boyfriend would be discussed, in relation to her behavioral and cognitive patterns within these relationships.

Biofeedback, relaxation exercises, and other cognitive-behavioral strategies might be utilized to control and reduce Nancy's anxiety level. A behavioral plan for exercise might also be created to help with her anxiety. Her thoughts about the meaning of her fears, about life and death, could be explored and restructured as would her thoughts about dependency, caretaking, money, and relationships. The major goals of this type of therapy would be to modify Nancy's disordered responses and help her develop alternative responses by restructuring her cognitive functioning and her actions so that she would be able to live more proactively and independently and, thus, feel better about herself. As Nancy learns to change her thinking and behavior, she will indeed change her emotional responses, and these changes will be reflected in her relationships and life choices.

Existential–Humanistic

All of the existential–humanistic theories share a phenomenological orientation, with their focus more on the subjective inner experience of people rather than on external objective forces or unconscious drives. Existential–humanistic approaches emphasize the worth of individuals and their unique meanings and approaches to problems. There is a strong emphasis on individual awareness, self-expression, and self-actualization. These approaches are concerned primarily with the authentic wholeness, passion, and uniqueness of the autonomous individual human being in the moment. The self is an active creator of its own reality and destiny—as a good, spontaneous, and free human being, rather than as a reactor to the demands of internal drives, other people, or society.

The common basic tenets of existential–humanistic theories are:

1. A phenomenological and experiential view of human behavior, where one's unique internal perspective and conscious experience determine one's reality.

2. A view that human beings have integrity as well as innate motivation for self-actualization, self-realization, and self-enhancement.
3. A belief in the essential freedom and autonomy of human beings, despite limits within human existence.
4. An anti-reductionistic point of view, in which experience is not reduced to basic drives, defenses, decontextualized behaviors, or thinking but is accepted for what it is.
5. An acceptance that human nature can never be fully defined.
6. An emphasis on the here and now, rather than on what was or what will be.

Rogerian (1951; 1961; 1980) client centered theory emphasizes the interaction of the self and the environment. If there is no congruence between the self-concept and the actual experience of the individual, mature full functioning will not develop and anxious, defensive functioning will result. Like object relations theory, this model proposes that a healthy self-concept requires empathic interactions with significant others. Rogers (1951) suggests that parents who are able to accept their child's feelings and strivings, who are able to accept their own feelings that certain of their child's behaviors are undesirable, and who are able to communicate their acceptance of the child as a person will enable him or her to develop a positive self-concept (self worth). Unlike object relations, however, current real relationships also shape an individual's experiences because past relationships are not thought to be internalized.

The goals of this model emphasize wholistic personal growth rather than specific symptom relief. Therapists aim to increase the congruency between feelings and behaviors, freedom of choice, emotional experiencing, and independent functioning. Experiential activities may be created within a trusting, empathic relationship to achieve these goals. Getting in touch with one's feelings and being able to communicate directly about them improves one's self-concept, awareness, and relationships.

Gestalt therapy, as promulgated by Perls (1969; 1972; 1976) and Polster and Polster (1973), focuses on powerful, directive techniques to reintegrate one's attention and awareness in order to integrate thinking and feeling and to take responsibility for actions. These techniques are particularly effective for people who lack awareness of the "how" and "what" of their present behavior and feelings; people who blame others and refuse to take responsibility for themselves and their lives; people who interact rigidly and in a ritualized manner with their environment; people who dwell on past unfinished business or on future rehearsing; and people who seem split in two because they deny or exclude part of themselves.

EXERCISE A.5 ▪ In small groups, identify the way you experience physically and express the following feelings: glad, mad, scared, and sad. What do you do with these feelings when you experience them? Whom can you tell about these feelings? Whom can you not tell?

EXERCISE A.6 ■ In pairs, take turns doing the gestalt two-chair dialogue. Think of some polarities in your temperament or disposition, that is, anxious versus calm; critical versus accepting; outgoing versus introverted. Using two chairs, be one of these characteristics in one chair talking to an imaginary you with the other characteristic in the other chair. Go back and forth between these two aspects of yourself arguing for your position in each chair. The purpose of this exercise is to increase awareness and achieve some kind of integration of both aspects.

Existential–Humanistic Conceptualization of Nancy The existential–humanistic therapist would focus primarily on the process of the therapeutic relationships, following whatever goals Nancy selected for therapy. In a genuine, empathic relationship with the therapist, Nancy would experience caring acceptance for the first time in her life, and this type of support would allow her to revise her suffering self-concept within this more positive relationship experience. In a more Rogerian client centered approach, the relationship, rather than techniques, would be the focus of existential–humanistic therapy: the therapist would avoid action, explanation, and emotional reassurances. Thus, there would be no exploration of Nancy's past experiences but instead a focus on her awareness of her here-and-now sensations and affective experiences. The aim would be to mobilize Nancy's inner strengths and resources so that she can experience the "educational component of love" (Havens, 1989, p. 303) from the therapeutic relationship. Once one has experienced a genuinely empathic, loving relationship, that experience is irreversible and will always be a part of one's life experiences.

In an approach that centralizes gestalt therapy, the therapist would use specific experiential enactments in a more structured approach to help her complete her "unfinished business." For example, the therapist might ask Nancy to sit in one chair as the needy, dependent Nancy and then have her respond from the other chair, as the stronger, achieving Nancy. The goal would be for Nancy to integrate her different parts and to gain understanding about the feelings, thoughts, and actions of each aspect of herself. Other exercises might include a dialogue between what Nancy experiences as the "good" mother and the "depriving" mother; again, the goal would be integration. The gestalt therapist might also direct Nancy to talk to herself in front of a mirror and exaggerate her facial behaviors. The therapist would continuously ask herself what she is feeling at this moment and help her to integrate inner and outer feelings.

Systems–Ecological Model

The systems–ecological model shifts from an intrapsychic focus to an interpersonal focus; from internal content to relational process; from reliance on an individual's report of relationships to observations of actual relationships and to a clearer understanding of how family, community, and larger sociocultural systems impact not only the individual, but also each other. Thus, contexts of the individual, the family, and the larger communities are essential data for understanding an individual's difficulties and goals for change. An individual cannot be considered outside the context of his or her relationships in family, at work, with peers, relatives, at school, in the neighborhood, and larger community and all of the other systems embedded in larger sociocultural systems.

It is assumed that difficulties stem from communicational and structural problems in the family or contextual systems. It is also assumed that the benefits and drawbacks of problematic (or healthy) thinking, feeling, or behaving apply not only to the individual, but also to all the people in the relational system. Psychological change is therefore challenging if only one person wants to change, because the other people will be acting in ways to encourage and therefore maintain the established problematic patterns. But a system is seen as interactional, so if a single person consistently changes his/her behavior, then others will have to change their behavior as well, and a new pattern will be established.

The primary mechanism of change in the systems-ecological model is to resolve the presenting problem(s) by altering the organization of the contexts (family or other). This is done by clarifying and changing patterns of communications; restructuring boundaries or the relational systems in order to clarify the hierarchy and distribution of power; identifying the ways that individuals support the system and each other in healthy or problematic ways; and changing the interactional patterns. Active, directive interventions that affect the relationships of the system (for example, the couple or family), rather than the person of the therapist or the relationship with the therapist, are the major emphasis of these approaches. Successful interventions require an accurate assessment of the problem in its context (that is, an appraisal of the system's contribution to presenting the problem and of the problem's effects on the system).

EXERCISE A.7 ■ Draw circles for each system in which you function: family of origin; current family; neighborhood; school; work; legal system; health system; religious groups; ethnic groups; gender groups; racial groups; friendship; and so forth. What is your role and function in each of these systems? How do they influence your identity and functioning?

EXERCISE A.8 ■ Consider how you affect and are affected by important people in your contexts. For example, when your partner or best friend is angry at something or someone other than you, how does this affect you? How do you react? Does it only affect you in the moment that you are with that person? When you are anxious or stressed, how do you act in ways that affect others? What do you expect others to do? What happens if they do or do not do what you expect? Discuss this in small groups and consider differences amongst you and your peers.

Systems–Ecological Conceptualization of Nancy A therapist with a systems-ecological model orientation would view Nancy's problems as emanating from within her family of origin system and larger sociocultural systems (school, community, work, and so forth). This therapist would focus on learning about the goals, rules, and roles of this family system as reflected in their communication patterns and boundary structures. Nancy's family might be viewed within different contexts and multi-generationally to surface the cultural and idiosyncratic themes, such as how attachment and separation issues with her mother may be related to ethnic and generational patterns in her mother's family. Nancy might be viewed as the "identified patient" who, by acting out the family pain, allows the rest of the family to pretend that nothing is wrong with anyone else in the family.

The systems therapist would most likely try to engage the rest of Nancy's family in the treatment. The goals would be to help the family system negotiate more effectively this launching phase of the family life cycle, to allow more differentiation for each family member, and to help the system get unstuck by improving family members' communications and restructuring "secret" coalitions. For example, Nancy's sister and parents seem to join together to blame her for any family conflict. If Nancy could feel less isolated and alienated from her family, she might be able to move on in her own life without having to resort demanding material things or engaging in sexual relations she does not enjoy as a way of seeking nurturance. If Nancy's family refused to participate, the systems therapist would actively coach Nancy in individual sessions to accomplish the same ends in her relationships with family members. The assumption is that if Nancy's interactional behavior with her family changes, their responses also will change, and a new cycle of interactional behavior will develop. As with cognitive-behavioral approaches, relationship behavior changes rather than insight would be the goal of therapy. Nancy's problems would be viewed in the context of her family system's dysfunction; her internal conflicts or character would not be the focus.

Constructivist Theories

Constructivism is more of an epistemological philosophy than a specific model of therapeutic orientation, but this philosophy shapes the ways in which theoretical orientations are understood and enacted. The basic tenets of constructivism are:

1. Individuals actively create their own worldviews in order to organize their experience. This means that there are multiple subjective realities that seem equally "valid" to the people holding a given particular view. To constructivists, whether an external reality exists is debatable and frequently largely irrelevant, as it is the subjective reality that affects experience.

2. People together (groups, societies) also create a shared worldview, a social reality. This socially constructed worldview is affected by and creates and

maintains social hierarchies and systems of power and privilege. Individuals' worldviews are created interactively with this social worldview, which thus constrains or enables options. Not all individuals have the same amount of influence on the shared, socially constructed social reality, and each individual is affected differently by that social reality.

3. The creation of one's reality is ongoing and continuously changing.

The constructivist therapist views people positively, respecting their "stories" or views of the world as emanating from their own meaning-making as well as societal meanings and assumptions. The belief is that people are active creators and construers of their realities and that they actively process information in order to explore and adapt to their environment. People are dynamic, decentralized structures, continually transforming and reorganizing their experiences in order to maintain order and coherence.

The constructivist philosophy has been most strongly embraced and formalized into theoretical orientation by cognitive-behavioral therapists, resulting in a variety of constructivist cognitive-behavioral approaches. These models are related to theories of information processing and emphasize the interdependence of thinking, feeling, and behavior. Emotions are valued as another way of knowing and are viewed as just as functional in the human information processing system (how people make meaning) as are cognitions. Constructivism focuses on the active and generative aspects of human cognitive processes and views the human mind as simultaneously proactive and reactive, with little distinction between input and output.

Constructivist therapists are concerned not only with the effects of the worldview on symptoms and behavior, but also with the development of the worldview as it affects the individual holistically. Anxiety and negative affect are viewed as necessary parts of development, not as something undesirable. This model integrates developmental, informational, and system theories. Constructivists frequently focus on the process whereby new information is incorporated into existing structures, thereby producing new structures, rather than on the rationalists' (traditional cognitive theorists) utilization of linear cognitive modification from the external world to the internal. As the environmental situations inevitably change, one is continuously remodeling and readapting, both to maintain and change one's cognitive organization and to be able to solve problems effectively.

According to constructivist views, psychopathology results from the discrepancy between environmental demands on the individual and that individual's current adaptive capacities and worldview (stories). In other words, psychopathology occurs when the necessary adaptive capacities for the process of cognitive reorganization are stuck or impaired. Individuals change as a result of changed self-knowledge, which provides new ways of thinking, feeling, and behaving. This adaptive cycle of change leads to more effective problem solving, which is necessary for growth and learning. Attachment is seen as the necessary foundation for the integration of maturational emotional and cognitive processes and environmental influences. The influence of emotional attachment on the development of self-knowledge and the regulating function of early experiences is significant.

In a constructivist approach, pathology is seen as related to the ways in which one's world view or story does not work in a current context or time, even though it may have previously. A child that grew up with emotionally abusive parents may, for example, develop a story that trusting others and becoming intimate is dangerous and hurtful and develop emotional and behavioral strategies of distancing and externalizing as protective. When that child grows up and has the opportunity for intimate relationships with people who are not emotionally abusive, his or her story about what relationships are like will no longer be functional. Alternatively, the functionality of a story may change not because the context changes from dysfunctional to functional (or vice versa), but because the context changes more generally and the shared social story changes. For example, if a family is relocated due to a parent's job transfer and the new community is very different than the previous community, the disconnect between one's view of self, others, and the world in the previous community may not be congruent with the cultural mores of the new community. This is common for immigrants and emigrants.

Attachment issues also frequently lead to a story that is not functional, as problems with attachment frequently result in distorted or negative self-knowledge, or challenges to development as when a failed separation impairs development by preventing the formal conceptualizations typical of adolescence to occur. These attachment disorders lead to distorted self-conceptions, or distortions in the "self-story," which are the causes of the kinds of dysfunctional thinking described by traditional cognitive theorists. When people feel stuck, they may need help in reconstructing their meanings and narratives.

Constructivist therapists provide a safe, caring, intense relationship and conduct a developmental, process-focused assessment and exploration of thinking, feeling and action. They utilize direct and indirect/relational emotional and behavioral interventions to foster insight into meaning of events and relationships as well as to change meanings more directly (through affect and behavior) without insight. Example techniques are: 1) reframing, turning a negative into a positive, such as telling a couple who comes because they are always fighting "But you're good at fighting and it shows how passionately you care about each other." which will change people's perspectives; and 2) cognitive restructuring, teaching new responses or asking "How else might you see this situation?" In other words, rather than judging people's thinking or behavior as rational or irrational, they collaboratively suggest other possibilities about how to perceive and interpret self, a situation, or other person.

Constructionism (Gergen, 2001), as opposed to constructivism, refers to the societal constructs that determine "social rules and norms." The interrelationship between sociocultural constructs and individual constructs (how one makes meaning of sociocultural constructs and constructs his/her own constructs) is an important part of therapeutic exploration.

EXERCISE A.9 ■ Frequently there are multiple stories about an event, depending on who is relating the event. Consider the following for example: Ben and Laura are going on their sixth date. After an evening at a nice restaurant and a movie (which Ben insisted on paying for), they return to Ben's dorm room. They begin to kiss as they have on other occasions, and both express that they really like each other. At some point, Ben begins to remove Laura's blouse and Laura protests that she is not sure she this is a good idea. Ben continues to unbutton her blouse however, and Laura does not immediately protest further. As they continue to kiss and fondle, Ben begins to unbutton Laura's pants, at which point Laura states that she does not think she wants to do this. Ben states that his roommate is gone all weekend, so there is no reason to worry and he continues to undo her pants. Laura abruptly sits up, fastens her pants and blouse, and says she needs to leave. Ben offers to walk her home, but she refuses. Ben calls her the next day to ask for another date, but Laura does not take his call that day, or in the several days that follow. What might Laura's narration of this night be like? What might Ben's narration be like? What might influence the meanings and interpretations they make? How do you understand what happened? What influences your story about this event? How do your own experiences, culture, and so on influence how you understand the event and the stories that are offered by those involved? What kinds of examples can you think of from your own life, where there may be very different perspectives on an experience? Laura and Ben likely had very different perspectives, but sometimes meanings can be only subtly different and still have significant implications.

EXERCISE A.10 ■ In small groups, discuss what the following mean to you and see if you can figure out what influenced your meaning-making: adultery; abortion; adoption; divorce; homosexuality; euthanasia; substance abuse; interracial marriage. The purpose of this exercise is to get in touch with some of your constructions and to compare them with others. How have sociocultural constructs (meanings) shaped your personal constructs? How would your meanings affect your therapy with a client who might be dealing with issues related to any of these constructs?

Constructivist Conceptualization of Nancy A constructivist therapist might view Nancy's difficulties as emanating from insecure attachment with her parents, resulting in distorted or negative self-knowledge and disorganization of thought processes (distorted thinking leading to distorted expectations.) A constructivist therapist would hear Nancy's story, her mother's story, and understand the influences of Nancy's peers and community on her beliefs (constructions) about life, particularly her materialism and expectations of others. Integrating cognitive, affective, and behavioral dimensions, the constructivist would help Nancy to "deconstruct" her assumptions, to consider alternative meanings, to broaden her expectations, and to learn to hear and understand different perspectives. This would be accomplished within a warm, collaborative relationship.

The style is exploration—seeking to understand stories and co-constructing new stories. The goal is adaptability and more effective problem solving.

Liberation Perspectives: Feminist/Multicultural

Like constructivism, liberation perspectives are based more in a philosophical viewpoint, than in a specific understanding of psychological development or therapeutic approach. A liberation philosophy emphasizes how the worldview and behavior of individuals are shaped by historical and current legacies of oppression and social power. Therapists with a primary liberation perspective are guided by the desire to address injustice and pain from oppression at both individual and social levels. Liberation perspectives were developed from the recognition that most psychological theories were developed in Western European or European American cultural contexts using white middle- to upper-class, heterosexual, male perspectives as the basis of health. Thus, many psychological theories have not been sensitive to the particular experiences of clients who are not in the dominant group, such as women, racial and ethnic minorities, GLBT, or economically disadvantaged clients. Furthermore, the lack of attention to the effect of relative powerlessness and oppression has led many theories to implicitly or explicitly pathologize individuals from these groups. Feminist and multicultural models specifically emphasize, respectively, on gender or on race and ethnicity as systems of oppression that affect the psychological functioning of individuals, while usually also examining intersections of oppressive systems.

Therapists with a primary feminist or multicultural orientation see pathology as primarily related to the oppressive systems that women and racial and ethnic minorities must negotiate. Active social change strategies as well as strategies to develop an individual's personal choice and autonomy are the major goals of feminist therapy. Feminist therapists emphasize egalitarian relationships with clients, and the therapist may frankly disclose her own values, expectations, beliefs, and attitudes in the service of this kind of relationship and also so that women clients understand that their problems are inseparable from society's oppression of women. Multicultural models emphasize the effects of cultural diversity, ethnocentrism, and racial discrimination and, like feminists, sensitize us to acknowledging and understanding powerlessness and oppression and to developing psychotherapeutic strategies that focus on the empowerment of clients rather than on their adjustment to mainstream society. They stress strategies for developing clients' competency skills, so that they may expand their opportunities, choices, and resistance to oppression.

The feminist and multicultural movements have challenged the patriarchal power structure and have insisted on the inclusion of diversity, pluralism, and interdisciplinary considerations in psychological theory. A liberation perspective, however, can be seen as broader than either of these more specific therapeutic orientations. Liberation perspectives focus generally on exploring the effects of social power and powerlessness, recognizing that therapy itself is a context of unequal power with a legacy of being oppressive to many types of people (women, cultural and racial minorities, GLBT people, economically disadvantaged people, religious minorities, and so on). Thus, therapists with a liberation perspective recognize that they must engage in significant self-examination in order to resist contributing to that legacy.

EXERCISE A.11 ▪ In small groups, share how your gender, ethnicity, race, sexual orientation, class, and geographical region have shaped your development. Pay particular attention to the intra-group differences in experiences and identities. How much awareness and knowledge do you have of people different from you?

EXERCISE A.12 ▪ How would you work with clients who you thought were being affected by discrimination (for example, sexism, racism, homophobia) who did not, themselves, believe that these issues were affecting their experience?

Liberation Perspective Conceptualization of Nancy A feminist/multicultural perspective with Nancy would focus on gendered cultural roles and the interdependence of personal and social identities. Sociocultural variables—race, gender, class, ethnicity, religion, sexual orientation, generation—give us clues to the norms of Nancy's family's closed, hierarchical family system and the transmitted notions about gender, sex, parenting, individuation, autonomy, closeness and distance, and achievement. Nancy's underlying feelings about herself as a female—for example, her need to shop for makeup and clothes and her perception that the only way to socialize is by drinking in bars—would be challenged and reframed as a form of oppressive cultural socialization within an egalitarian therapeutic relationship. Her notions of privilege and her assumptions about affluence in her community would be thoroughly explored. The goal would be to enable Nancy to have equal-in-power relationships with both females and males, to accept her own body image and different worldviews so that she is not pathologized or oppressed by dominant cultural norms.

A competent therapist oriented to any of these theoretical models could probably help Nancy. Although the models have different and often conflicting viewpoints about human nature and development, health and pathology, the process and content of psychotherapy, the role of the therapist, the focus of treatment, and the mechanisms of change, each has an important perspective and focus. The point is that the therapist's choice of theoretical conceptualization for a given client depends on the fit between the theory and a) the therapist's personal variables; b) the client's particular experiences; and c) the therapist's values and views about change, development, pathology, and so on. As discussed, theories are not static. Practitioners and subsequent generations of theorists continuously refine theories as they incorporate new findings and expanded paradigms.

REFERENCES AND RECOMMENDED READINGS

Adler, A. (1925). *The practice and theory of individual psychology* (trans. by P. Radin) (Rev. ed.). London: Routledge & Kegan.

Beck, A.T. (1967). *Depression: Clinical, Experimental, and Theoretical Aspects*. New York, NY: Harper & Row.

Beck, A.T. (1976). *Cognitive theory and the emotional disorders*. New York, NY: International Universities Press.

Berman, S. (1997). *Children's social consciousness and the development of social responsibility*. Albany, NY: State University of New York Press.

Eells, J.E. (2006) (Ed.). *Handbook of psychotherapy case formulation* (2nd ed.). New York, NY: Guilford.

Freud, S. (1936). *The problem of anxiety*. New York, NY: Norton.

Freud, S. (1943). *A general introduction to psychoanalysis* (1915–1917). New York: Garden City.

Freud, S. (1953). *The interpretation of dreams*. Standard Edition, 4 & 5. London: Hogarth Press.

Freud, S. (1964). *An outline of psycho-analysis*. Standard Edition 23, 141–207. London: Hogarth Press.

Gergen, K.J. (2001). *Social construction in context* (2nd ed.). London: Sage.

Gulyas, B. (2009). Functional neuroimaging and the logic of conscious and unconscious mental processes. *Neurological Correlates of Thinking* 1(11), 14–173.

Havens, L. (1989). Clinical interview with a suicidal patient. Suicide: understanding and responding. *Harvard Medical School Perspectives,* 343.

Hersen, M. & L.K. Porzelius (2002). *Diagnosis, conceptualization, and treatment planning for adults: A step-by-step guide*. New York, NY: Lawrence Erlbaum.

Kohut, H. (1971). *The analysis of the self: A systematic approach to the psychoanalytic treatment of narcissistic personality disorders*. New York, NY: International Universities Press.

Kohut, H. (1977). *The restoration of the self: A systematic approach to the psychoanalytic treatment of narcissistic personality disorder*. New York, NY: International Universities Press.

Kuhn, T.S. (1962). *The structure of scientific revolutions* (2nd. ed.). Chicago: University of Chicago.

Norcross, J.C. & Prochaska, J.O. (1983). Clinicians' theoretical orientations: Selection, utilization, and efficacy. *Professional Psychology: Research and Practice* 14, 197–208.

Okun, B.F. (1990). *Seeking connections in psychotherapy*. San Francisco: Jossey-Bass.

Perls, F. (1976). *The Gestalt approach and eye witness to therapy*. New York, NY: Bantam.

Perls, F.S. (1969). *Ego, hunger, and aggression: The beginning of gestalt therapy*. New York, NY: Random House.

Perls, F.S. (1972). *Gestalt therapy now*. New York, NY: Harper & Row.

Polster, E. & M. Polster (1973). *Gestalt therapy integrated: Contours of theory and practice*. New York, NY: Brunner/Mazel.

Rogers, C.R. (1951). *Client-centered therapy: Its current practice, implications and theory*. London: Constable.

Rogers, C.R. (1961). *On becoming a person*. Boston: Houghton Mifflin.

Shevrin, H., Bond, J.A., Brakel, L.A.W., Hertel, R.R., & Williams, W.J. (1996). *Conscious and unconscious processes: Psychodynamic, cognitive and neurophysiological convergers*. New York, NY: Guilford.

Stevens, M.J. & S.J. Morris (1995). A Format for Case Conceptualization. *Counselor Education and Supervision* 35(1), 82–94.

Wampold, B.E. (2001). *The great psychotherapy debate: Models, methods, and findings*. Mahwah, NJ: Erlbaum.

Appendix B

Exploring Your Experiences with Culture, Power, and Privilege

In this appendix, we *briefly* review some of the issues related to culture, power, and privilege that will affect your case conceptualization, treatment planning, and approach to therapy. Our intention here is simply to remind you of issues that you have encountered in your prior training or, if you have not, to help you identify areas for further exploration and resources that will help you in this task.

BRIEF REVIEW OF EXPLORING YOUR ETHNIC CULTURE

As you likely know from your earlier training, our values and experiences—including our ideas about health, pathology, and change—are cultured, that is, affected by the values, norms, and behavioral patterns of the ethnic culture of the systems (family, ethnic group, society, and so forth) in which we have developed. Research on culture has identified many areas of cultural difference that affect how we understand and interact with each other. These include:

- *The nature of human beings and their influence on environment.* This may include whether people have inherent morality, characteristics, or "drives"; whether there is a distinction between the mind and the body; the nature of the relationship between people and the environment;

perceptions of our ability to influence our own experiences and our environment (including our ability to create change!); and values about the best ways to address problems (including the value of insight into the causes or effects of those problems).

- *Communication styles*. This may include how much individuals value verbal expression and expressiveness; the meaning of honesty; the meanings of silence in communication; whether the meaning of communication is in the content of the words spoken or the context of the speakers and situation (*low vs. high context communication*); the meaning and commonness of nonverbal communication such as eye contact, body movement (*kinesics*), and personal space (*proxemics*); and norms for verbalization such as rate of speech, loudness, quickness in responding.

- *The nature of self and relationships*. This may include attitudes towards authority and values of hierarchy or egalitarianism in relationships; values about autonomy, individual achievement, and competition versus interdependence, reciprocal obligation, and cooperative effort (related to *individualism* vs. *collectivism*, and *independent* vs. *interdependent self-construal*); criteria for trustworthiness or interpersonal credibility; and the appropriateness of emotional expression and self-disclosure in relationships.

As a therapist, it is imperative that you are aware of your own cultured values and worldview, so you may understand how these affect your understanding of and approach to your clients. You have likely explored this in your earlier training, so we simply provide here some exercises to remind you of your own positions and views.

EXERCISE B.1 ▪ Below are a number of statements about the nature of people. Choose where you would place yourself on each continuum. What do you believe about people?

People are inherently good	People are inherently neutral	People are inherently bad
People are born with the capacity to change		People are born with inherent characteristics that cannot change
People make their own luck or limitations		People are influenced by fate/God
People have an inherent tendency towards self-actualization or healthy functioning		Left alone, people who have developed problematic ways of being will not change
People have an inherent "drive" towards sexual gratification		There are no inherent drives, just choices
People have an inherent "drive" towards relationships with others		There are no inherent drives, just choices
People are no different than other living things (animals, plants, the planet)		People are inherently different from other living things
People are meant to master the environment		People are meant to protect and care for the environment
People should recognize the distinction between mind and body		People should see the mind and body as one thing

People can change almost anything if they want to	People are primarily determined by genes and circumstances
It is important to express one's thoughts and feelings	It is better to not impose one's thoughts and feelings on others
Complete honesty is important to me	Sometimes it is better to be silent than to say what you think
I am silent primarily when I disagree with what is said but do not want to be disrespectful	I am silent primarily when I agree or have nothing more to add
I am comfortable when people respond quickly and we debate	When people respond quickly, I feel they are not really listening
I believe what people say, the words themselves	I believe people do not always say what they mean directly
People need to understand themselves in order to change	People should just "do it" (change the way they act)
Relationships have a natural hierarchical order that should be respected	Egalitarian relationships are the best
Everyone has equal opportunity	People are limited by circumstances they cannot control
People should strive to achieve as much as they can	People should compromise their personal goals to excel for the group's/family's advancement
I trust others who know more than I do	I trust others in authority
Individuals should make their own decisions	Individuals should make decisions based on the influence and needs of others
People make their own fate	God determines one's fate
People should question those in authority	Authority should be respected
I am suspicious of those who do not make eye contact when I am speaking	I believe direct eye contact is disrespectful to the speaker
Rational communication is best	Without emotional passion, communication is meaningless
People should think independently	It is best to be guided by the wisdom of others, such as family
I would rather win a prize individually	I would rather win a prize as a member of a group
I am influenced a lot by what others think and need	I make my own decisions
I do not like to rely on others	I am most comfortable depending on others and having them depend on me
I like to tell people about myself and my family	I am a very private person
When I have a problem, I work it out for myself	When I have a problem, I talk about it with my friends and colleagues

EXERCISE B.2 ■ What are the cultural and personal experiences that have shaped your ratings in the previous exercise? How have your cultural and personal experiences shaped your comfort level with different ways of thinking and approaching interpersonal (and therapeutic) relationships?

Culturally competent therapy involves not only knowing about your own cultured experiences, but also the experiences of individuals and groups from cultures different than yours (and/or different from the dominant culture in your context). Most initial training for counselors and therapists involves explorations of the experiences of diverse cultural groups.

EXERCISE B.3 ▪ As a review, consider what you know about the modal experience of different cultural groups and their views on the items listed in Exercise B.1. Discuss with your peers to try and pool your knowledge and memory. As an alternative (or in addition), discuss the items with as many people from different cultural groups as you can or research different groups. What kinds of difficulties might you encounter when interacting with someone who answers these questions differently than you do?

BRIEF REVIEW OF EXPLORING YOUR POWER AND PRIVILEGE

As with culture, your earlier training has likely involved an exploration of how systems related to power and privilege (for example, gender, race, sexual orientation, social class) have affected you. As we said in Chapter 2 (page 22),

> By *power*, we mean the ability to influence circumstances (and people!) in a desired direction. By *privilege*, we mean having preferred status in a social system of hierarchy that benefits some, but not others in ways that are not connected to effort or ability. People in privileged spaces have more power because they not only have the power that they earn/create, but also the power that is given to them (and not to others) because of their status.

McIntosh (1988) describes White privilege as "an invisible package of unearned assets that I can count on cashing in each day, but about which I was 'meant' to remain oblivious." She notes that privilege and oppression exist in:

> active forms, which we can see, and embedded forms, which as a member of the dominant groups one is taught not to see. In my class and place, I did not see myself as a racist because I was taught to recognize racism only in individual acts of meanness by members of my group, never in invisible systems conferring unsought racial dominance on my group from birth.

The important point here is that it is hard to see the effects of being privileged in relation to these systems because we are socialized to believe that the power and benefits we have are because we earned them. In addition, we frequently do not see how everyday experiences are related to systems of power and privilege (McIntosh, 1988; Goodman, 2001; Pinderhughes, 1989; Sue & Sue, 2007).

EXERCISE B.4 ■ Answer the questions below. After having completed your answers, consider what you have done to contribute to having that experience. Consider also, what the experience of someone who answers differently than you might be like.

1. Do you consider your physical safety when you go out on a date with a new person? What are your concerns?
2. Were the history and accomplishments of people of your gender regularly taught to you in school? Who were the people you learned about?
3. Do you regularly see symbols of your major religious holidays or cultural practices in public spaces and shops at appropriate times of the year?
4. Are your religious holidays and practices generally known and understood by others?
5. Is your place of work or school closed on your major religious holidays?
6. Can you openly show affection for your romantic partner in most settings without fear? Is there variability across settings and, if so, what contributes to this variability?
7. Can you display a picture of your significant other at work without worry or comment from others?
8. Can you marry your current or most recent partner if you wanted to and have that marriage recognized legally and socially anywhere you go?
9. Were you raised in a community where the majority of elected officials, teachers, or police were of your racial and/or ethnic group?
10. Have you ever been stopped/harassed by the police or followed in a store by clerks because of your race or ethnicity?
11. Have you had to make considerable effort to buy greeting cards that depict people who look like you?
12. Have you had negative names, words, or sounds about your group membership called out to you in public from strangers?
13. Do you usually see people like you and relationships similar to yours portrayed positively in movies, TV shows, and magazines?
14. Have you had frequent moments considering whether something happened because of discrimination or because of your personal behavior?
15. Were you or your ancestors ever legally denied full legal rights and privileges because of their categorization or status, rather than because of something you or they did?

While some trainees may be familiar with questions such as these, many have not fully considered how these experiences affect their understandings of how people change, what constraints exist in relation to change, or how they interact with clients who have different experiences with power and privilege. The areas that you have not considered fully are most likely to be areas in which you have systemic privilege related, for example, to gender (questions 1–2), religion (questions 3–5), sexual orientation (questions 6–8), or race (questions 9–11). The last group of questions is related to experiences that may relate to multiple areas of privilege. We included a historical question (#15) because we recognize that historical legacies still affect current experiences and that these experiences affect interactions in psychotherapy (see Vasquez & McGraw, 2005). If your grandparents were unable to enroll in college, that can affect the educational and economic opportunities available to your parents, and thus to you. If your grandparents were imprisoned in Japanese American concentration camps or Nazi death camps during World War II, this can affect your family dynamics, values, your own sense of self-esteem and safety, and your attitudes towards authority (Bar-on, 1995; Bowen, 1978; Nagata, 1993). These are, of course, just a few examples.

If you have not yet fully explored your own cultural background and experiences of power, privilege, and relative oppression, we strongly encourage you to do so. The References and Resources list at the end of this chapter provides resources to further this exploration or help you in your review of these issues.

REFERENCES AND RESOURCES

Adams, M.W.J., Blumenfeld, R., Castañeda, R., Hackman, H.W., Peters, M.L., & Zúñiga, X. (Eds.). (2000). *Readings for diversity and social justice: An anthology on racism, antisemitism, sexism, ableism, and classism*. New York, NY: Routledge.

American Psychological Association. (2003). Guidelines on multicultural education, training, research, practice, and organizational change for psychologists, *American Psychologist* 58, 377–402.

APA Council of Representatives. (2011). *Guidelines for psychological practice with lesbian, gay, and bisexual clients*. Washington, D.C.: American Psychological Association.

Anderson, S.K., & Middleton, V.A. (Eds.). *Explorations in privilege, oppression, and diversity*. Belmont, CA: Brooks/Cole.

Bar-on, D. (1995). *Fear and hope: Three generations of the holocaust*. Cambridge, MA: Harvard University Press.

Bem, S.L. (2004). Transforming the debate on sexual inequality: From biological difference to institutionalized androcentrism. In J.C. Chrisler, C. Golden, & P.D. Rozee (Eds.), *Lectures on the psychology of women* (3rd ed.), 3–15. Boston: McGraw Hill.

Bowen, M. (1978). *Family therapy in clinical practice*. New York, NY: Aronson.

Garnets, L. D. (2002). Sexual orientations in perspective. *Cultural Diversity and Mental Health*, 115–129. New York, NY: Columbia University Press.

Glick, P., & Fiske, S.T. (2001). An ambivalent alliance: Hostile and benevolent sexism as complementary justifications for gender inequality. *American Psychologist* 56, 109–118.

Golden, C. (2004). The intersexed and the transgendered; Rethinking sex/gender. In J.C. Chrisler, C. Golden, & P.D. Rozee (Eds.), *Lectures on the psychology of women* (3rd ed.), 95–109. Boston: McGraw Hill.

Goodman, D. J. (2001). Promoting diversity and social justice: Educating people from privileged groups. Thousand Oaks, CA: Sage.

Kliman, J. (2005). Many differences, Many voices: Toward social justice in family therapy. In M. Mirkin, K.L. Suyemoto, & B. Okun (Eds.), *Psychotherapy with Women: Exploring Diverse Contexts and Identities.* New York, NY: Guilford Press.

Liu, W. M., Soleck, G., Hopps, J., Dunston, K., & Pickett Jr., T. (2004). A new framework to understand social class in counseling: The social class worldview model and modern classism. *Journal of Multicultural Counseling & Development* 32, 95–122.

Lott, B. & Bullock, H.E. (2001). Who are the poor? *Journal of Social Issues* 57, 189–206.

MacIntosh, P. (1988). *White privilege and male privilege.* Working paper 189. Wellesley, MA: Wellesley College Center for Research on Women.

Nagata, D.K. (1993). *Legacy of injustice: Exploring the cross-generational impact of the Japanese American internment.* New York, NY: Plenum Press.

Pinderhughes, E. (1989). *Understanding race, ethnicity and power: The key to efficacy in clinical practice.* New York, NY: Free Press.

Pope-Davis, D.B., Coleman, H.L.K., Liu, W. M., & Toporek, R.L. (2003). *Handbook of multicultural competencies in counseling and psychology.* Thousand Oaks, CA: Sage Publications.

Smedley, A. & Smedley, B.D. (2005). Race as biology is fiction, racism as a social problem is real: Anthropological and historical perspectives on the social construction of race. *American Psychologist* 60, 16–26.

Sue, D.W., & Sue, D. (2007). *Counseling the culturally diverse: Theory and practice* (5th ed.). New York, NY: Wiley.

Suyemoto, K.L., & Dimas, J.M. (2003). To be included in the multicultural discussion: Check one box only. In J.S. Mio & G.Y. Iwamasa (Eds.), *Multicultural mental health research and resistance: Continuing challenges of the new millennium.* New York, NY: Brunner-Routledge.

Toporek, R. L., Gerstein, L., Fouad, N., Roysicar, G., & Israel, T. (2006). *Handbook for social justice in counseling psychology: Leadership, vision, and action.* Thousand Oaks, CA: Sage Publications.

Vasquez, H., & McGraw, S. (2005). Building relationships across privilege: Becoming an ally in the therapeutic relationship. In M. Mirkin, K.L. Suyemoto, & B. Okun (Eds.), *Psychotherapy with women: Exploring diverse contexts and identities.* New York, NY: Guilford Press.

Weber, L. (1998). A conceptual framework for understanding race, class, gender, and sexuality. *Psychology of Women Quarterly* 22, 13–32.

Appendix C

Nancy's Genogram

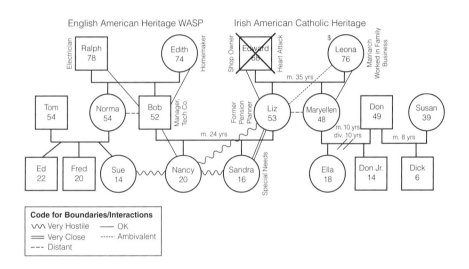

Index

Page numbers followed by *f* indicate figures; and those followed by *t* indicate tables.

A

Addressing questions, skills in conceptualization and, 1
Advocacy for evidence based practice, 9
Affect and cognition, Nancy's therapist continues to focus more on, 187
Affective change frequently most difficult for novice therapists, 59
American Psychological Association (APA), Task Force on Evidence Based Practice from, 9
Approach to clients, therapists intuitively modify their, 140
Approaches, exploring ramification of different, 117
Arguments with parents, Nancy had heated, 183
Assessments
  economic constraints on obtaining formal, 144
  in English, 128
  referring client for formal, 144
Assessments and other consultation, formal, 143
Assumptions are conveyed directly in language, some, 132
Attachment theory, 156
Attention, Nancy to have therapist's support and, 216
Attention Deficit Disorder, 201

B

Behavior, focus on, 58
Beliefs about change, understanding, 19–21
Beliefs being absolute, values and, 104
Bilingual clients, language issues may also affect, 128
Billing or organizational reasons, assigning diagnosis to your client for, 173
Biofeedback and relaxation training, 133
Body, clients respond facially or in their, 138
Boredom, therapist's, 140

C

Case conceptualization
  and developing effective treatment plan, 147
  founded upon theoretical orientation, 147
  is applied theoretical orientation, 6
Case conceptualization knowledge and skills, ways of teaching, 148
Case conceptualization, three main components of, 4
  clinical formulation, 4
  diagnostic formulation, 4
  treatment formulation, 4
Change
  clients seems stuck or resistant to, 73
  dimension and established theories, 61
  relationship with therapist and, 81
  therapeutic interventions contribute to positive, 182
  understanding beliefs about, 19–21
  understanding process of, 2
Change, influences on ideas about, 21–25
  beliefs about people and worldview, 21–22
  cultural influences on understanding clients, 21–22
  power, privilege, and inequity, 22–24
  power and privilege related to role of therapist, 24–25
Change in therapy, preference for facilitating, 34*t*
Chapters contain many exercises, 3
Characteristics, client's overt, 126
Choices, making about what is central to understanding, 147
Client
  characteristics contributing to comfort or discomfort of therapists, 100
  conceptualization becomes embedded in ongoing story about, 185
  discomfort or even criticism of therapist, 215

258

family developmental stage for given, 46
fit of therapist and, 98–107
information from observations of and interactions with, 137
most important thing to convey to, 137
not happy with something therapist does or says, 215
Client about client contexts, information from, 131
Client about presenting problem, information from, 130
Client for billing or organizational reasons, assigning diagnosis to your, 173
Client for formal assessment, referring, 144
Client from observation, Nancy's therapist gathered information about, 152
Clients
  gathering information from, 137
  goals, 162–163
  inquiring about sexual orientation, 112
  inquiry maybe matter of social convention, 112
  language issues may also affect bilingual, 128
  needs, 64
  not every clinician can be effective with all, 107
  overt characteristics, 126
  respond facially or in their body, 138
  seems stuck or resistant to change, 73
  sharing information obtained from, 143
  terminations are frequently difficult for therapists as well as, 220
  therapists not liking, 107
  with unrelated problems, 105
  verbalizations from, 135
Client's contexts, understanding, 133
Client's direct expressions, information from, 127, 136
Clients enter room, facial expressions when, 138
Client's experience, understanding, 130
Client's functioning, hypothesis about, 126
Clients in contexts, conceptualizing, 39–65
  dimensions related to conceptual understandings, 52–64
    context location of problem, 53–58
    focus on change, 58–64
  exploring contexts, 39–52
    exploring family contexts, 44–46
    exploring sociocultural and sociostructural systems, 46–52
Client's multiple contexts, skill and knowledge related to, 103
Clients' own evaluations of progress, therapists incorporate, 211
Clients present as information, therapy and distinction between what, 182
Client's story, individual, 45
Clients struggles, therapists and, 141
Clinical formulation, 4
Clinician can be effective with all clients, not every, 107
Cognition, Nancy's therapist continues to focus more on affect and, 187
Cognitive-behavioral theory, 156
Collateral information, 142–143
Comfort or discomfort of therapists, client characteristics contributing to, 100
Communicating reasons for referral, 107
Communication, process, 89
Communication processes, learning about and understanding, 44
Community events and resources, searching for regional, 49
Conceptualization
  becomes embedded in ongoing story about client, 185
  diagnosis can contribute to narrowing, 172
  difficult to describe process of continual, 180
  is application of theoretical orientation, 5–6
  is continuous, 209
  is relational process, 7
Conceptualization, beginning, 125–149
  beginning process of case conceptualization, 125–145
  connecting information to theory, 146–148
  gathering information, 125–145
    formal assessments and other consultation, 143
    information from client's direct expressions, 127
    information from observations of and interactions with client, 137
    primary sources of information, 127–142
    secondary sources of information, 142–145
  moving towards conceptualization, 146–148
  organizing and integrating information (pre-conceptualization), 145–146
Conceptualization, exploring, 1–15
  conceptualization defined, 4–10
    conceptualization and theoretical orientation, 5–6
    conceptualization is continuous process, 10
    conceptualization is integral part of evidence based practice, 7–9
    conceptualization is relational process, 6–7
  conceptualizing Nancy, 13–14
  foundational understandings about people and learning, 2
  introducing case of Nancy, 10–13
    attitude towards therapy, 13
    background, 11–12
    presenting concerns and strengths, 12
    referral source, 10
    verbal and non-verbal behavior, 12–13
Conceptualization, iterative and treatment planning, 179–207
  case of Juan, 194–203
    information from client from first five sessions, 194–196
    information from observations of client and counselor-client interactions, 196
    initial conceptualization and treatment, 197–199
    initial information, 194–197
    iterative conceptualization and treatment, 199–203
  continual conceptualization and integrating new information, 183–188
    revisiting organizing information for continual conceptualization, 184
    testing hypotheses from prior information and conceptualizations, 184–188
  continual conceptualization with Nancy, 181–183
  encountering surprises, 188–194
  interactions with systems of care, 206
  iterative conceptualization, consultation, and adjunct treatments, 204–205
  iterative conceptualization from start to finish, case of Juan, 194–203
    information from client from first five sessions, 194–196
    information from observations of client and counselor-client interactions, 196
    initial conceptualization and treatment, 197–199
    initial information, 194–197
    iterative conceptualization and treatment, 199–203
  and modifying diagnoses, 203–204
  ongoing records and clinical notes, 205–206

Conceptualization, iterative and treatment planning (*continued*)
   process of iterative conceptualization, 188–194
Conceptualization, theoretical orientation is foundation for, 6
Conceptualization, treatment planning, and diagnosis, 151–177
   choosing strategies for treatment goals, 169–171
   conceptualization and treatment planning, 158–169
      developing goals, 162–169
      therapists' intervention preferences, 158–160
      treatment planning - journey, not a destination, 160–162
   diagnosis, 171–175
   formalizing conceptualizations and treatment plans, 175–176
   information gathering with early conceptualization and treatment planning, 156–158
      initial treatment planning during information gathering with Nancy, 157–158
      integrating information and moving towards conceptualization with Nancy, 151–156
Conceptualization and addressing questions, skills in, 1
Conceptualization and treatment, cyclical process of, 188
Conceptualization not the same, theoretical orientation and, 5
Conceptualization regarding Nancy, organizing information necessary for, 153
Conceptualization with Nancy, integrating information and moving towards, 151–156
Conceptualizing, involves relational style as therapist, 67
Conceptualizing clients in contexts, 39–65
   dimensions related to conceptual understandings, 52–64
      context location of problem, 53–58
      focus on change, 58–64
   exploring contexts, 39–52
      exploring family contexts, 44–46
      exploring sociocultural and sociostructural systems, 46–52
Conceptualizing Nancy, therapists initial thoughts about, 153
   challenges, 155–156
   how client's problems serve client, 155
   Nancy's perception of her distress, 153
   strengths and resources, 155
   therapist's conceptualization of contextual influences, 154–155
   therapist's views of Nancy's most pressing problems, 153–154
Conceptualizing therapeutic relationships, 67–96
   pulling it all together, 91–95
   relationships and location of knowledge, 90–91
   role of relationship in contributing to change, 76–90
      process emphasis, 87–90
      process emphasis dimension to established theories, 89
      real unreal relationship, 82–87
      real-unreal relationship dimension to establish theories, 86
      therapeutic relationship, 79–82
      therapeutic relationship dimension to established theories, 81–82
      working alliance, 77–79
   understanding relational style, 68–76
      confrontativeness, 72–74
      dimensions of relational style to established theories, 74–75, 75*t*
      directive versus non-directive, 68–70
      level of activeness, 71–72
      relational style dimensions and role of therapist, 75–76
      structuring process, 70–71
Confidentiality, managed care and, 116
Confrontation is usually uncomfortable, 73
Connections, making, 145
Constraining
   diagnosis can be helpful or, 172
   effects, 61
Constraints on goals, 164–167
Consultation
   formal assessments and other, 143
   on-going peer supervision or, 121
   and supervision, 121
Context far away from family, cultural, 134
Context of psychotherapy, social, 26–33
Contexts
   affecting development of psychological experience, 53
   cultural, 46
   ecological, 41*t*, 42*f*
   emphasis on dimension and established theories, 57
   environmental, 40
   exploring, 131
   family systems, 60
   of given therapy, 136
   information from client about client, 131
   learning about particular, 48
   referral, 126
   role, rules, function, and responsibilities, 43
   skill and knowledge related to client's, 103
   understanding client's, 133
   understanding different, 39
Contexts, conceptualizing clients in, 39–65
   dimensions related to conceptual understandings, 52–64
      context location of problem, 53–58
      focus on change, 58–64
   exploring contexts, 39–52
      exploring family contexts, 44–46
      exploring sociocultural and sociostructural systems, 46–52
Contextual factors, positive, 167
Contextual issues, knowledge about larger, 134
Continual conceptualization
   defined, 183
   difficult to describe process of, 180
Continuous, conceptualization is, 209
Continuous evaluation and termination, 209–231
   continuing with new goals or to end therapy, 218–219
   continuous evaluation of clients and their progress in therapy, 210–218
      assessing during session, 213
      assessing what happens between sessions, 217
      assessing what happens over time, 217–218
      continuous evaluation of effectiveness as therapists, 211–212
      evaluating client progress, 212–218
   final thoughts, 229–231
   follow-up and returning clients, 228–229
      follow-up, 228–229
      returning clients, 229

termination, 219–228
    finances and termination, 223
    planned termination, 223–226
    termination with Nancy, 227–228
    unplanned termination, 221–222
Continuum, directive/non-directive, 70
Counselor-client interactions, information from, 139
Criticism of therapist
    client discomfort or even, 215
    differences between discomfort and, 216
Cultural context far away from family, 134
Cultural contexts, 46
Cultural diversity of different groups, 49
Cultural experiences, relational norms, expectations, values and, 78
Cultural groups, knowledge of modal experiences of, 50
Cultural meaning and affects of sociostructural system, 51
Cultural or personal differences in expectations, 78
Cultural sensitivity, social justice and, 116–121
    clients' bias, 120–121
    clients' internalized oppression, 118–120
    discrimination related to therapist minority status, 120–121
Culture, knowing about general importance of, 47
Cyclical process, 180f
Cyclical process of conceptualization and treatment, 188

D

Decoding meanings, 145
Definition of disorders, 9
Depression, 130
Derailing therapy, 213
Developing skills in conceptualization and addressing questions, 1
Development, ecological model of human, 2
Diagnoses
    can be helpful or constraining, 172
    can contribute to narrowing conceptualization, 172
    classification system of, 171
    concept of, 171
    DSM five-axial, 174
    validates that there is a problem, 172
    variety of, 173
Diagnoses, conceptualization, treatment planning, and, 151–177
    choosing strategies for treatment goals, 169–171
    conceptualization and treatment planning, 158–169
        developing goals, 162–169
        therapists' intervention preferences, 158–160
        treatment planning, journey, not a destination, 160–162
    diagnosis, 171–175
    formalizing conceptualizations and treatment plans, 175–176
    information gathering with early conceptualization and treatment planning, 156–158
        initial treatment planning during information gathering with Nancy, 157–158
        integrating information and moving towards conceptualization with Nancy, 151–156
Diagnosis to your client for billing or organizational reasons, assigning, 173
Diagnostic and Statistical Manual (DSM), 9, 30–31
Diagnostic and Statistical Manual (DSM) five-axial diagnoses, 174
    deferred, 174

Generalized Anxiety Disorder (GAD), 174
Global Assessment Functioning is overall view of daily performance, 174
Nancy did not get attention from her parents when sick, 174
numerous factors contributing to Nancy's anxieties, 174
Diagnostic formulation, 4
Differences, psychotherapy and ideological, 104
Dilemmas
    in psychotherapy and ideological differences, 104
    regarding non-payment of fees, 115
    that may emerge in therapy, 97
Dilemmas about disclosure, therapists faced with, 113–114
Dilemmas in effective helping, 97–123
    boundary dilemmas, 108–116
        boundary issues related to managed care systems, 115–116
        disclosure, 110–115
    consultation and supervision, 121
    fit of therapist and client, 98–107
    social justice and cultural sensitivity, 116–121
        clients' bias, 120–121
        clients' internalized oppression, 118–120
        discrimination related to therapist minority status, 120–121
Dimensional preferences, 129
    therapist influenced by theoretical orientations and, 187
Dimensions
    of focus, 34t
    preferences on relational, 142
Dimensions to traditional theoretical orientations, relation of, 36t
Directive therapists, 69
Directive/non-directive continuum, 70
Disclosing something from very beginning of therapy, 113
Disclosure
    and online social networking and online images, 114
    therapists faced with dilemmas about, 113–114
Discomfort and criticism of therapist, differences between, 216
Discomfort of therapists, client characteristics contributing to comfort or, 100
Discomfort or even criticism of therapist, client, 215
Disorders, definition of, 9
Diverse people, foster relations with, 49
Diversity of human experience, 3
Door knob phenomenon, 215
Drinking was not a problem, Nancy insisted that, 189
DSM. *See* Diagnostic and Statistical Manual (DSM)

E

Eating disorder program, Nancy investigated intense outpatient, 193
Ecological contexts and influences, 41t, 42f
Ecological model of human development, 2
Economic constraints on obtaining formal assessments, 144
Effective, novice therapists are asking themselves if they are being, 211
Effective helping, dilemmas in, 97–123
    boundary dilemmas, 108–116
        boundary issues related to managed care systems, 115–116

disclosure, 110–115
consultation and supervision, 121
fit of therapist and client, 98–107
social justice and cultural sensitivity, 116–121
clients' bias, 120–121
clients' internalized oppression, 118–120
discrimination related to therapist minority status, 120–121
Effectiveness within session, 214
Emotion-focused therapy, 82
Empirically Supported Therapies (ESTs), 7–8
Empirically Validated Therapies (EVTs), 7–8
English, assessment in, 128
Environmental context, 40
Established theories have some structure to them, 75
Ethical mandates, 105
Ethnic culture is sociostructural system, 51
Evaluate, outcomes therapists, 212
Evaluating and using information from interactions, challenge in, 140
Evaluation and termination, continuous, 209–231
continuing with new goals or to end therapy, 218–219
continuous evaluation of clients and their progress in therapy, 210–218
final thoughts, 229–231
follow-up and returning clients, 228–229
termination, 219–228
finances and termination, 223
planned termination, 223–226
termination with Nancy, 227–228
unplanned termination, 221–222
Evaluation of given session, 214
Events and resources, searching for regional community, 49
Evidence Based Therapies, 7
Exercises, chapters contain many, 3
Expectations, cultural or personal differences in, 78
Experiences
contexts affecting development of psychological, 53
diversity of human, 3
learning is a mix of knowledge, reflection, and, 3
relational norms, expectations, values and cultural, 78
strategies for increasing knowledge and, 48–49
understanding client's, 130
Experiences, multiple types of, 40
environmental context, 40
identities, 40
ideologies, 40
relationships that people have, 40
Experiences of cultural groups, knowledge of modal, 50
Expressions, information from client's direct, 127, 136

F

Facial expressions when clients enter room, 138
Facially or in their body, clients respond, 138
Family
cultural context far away from, 134
developmental stage for given client, 46
of origin, 134
Family, social developmental challenges and attitudes, 209
Family systems
context, 60
theories, 57, 156
Family therapy session, Nancy and, 152
Fear, trainee's, 101

Fears and reactions, Nancy is more willing to share, 186
Feelings
and interactions, 83
and thoughts guiding self-evaluation as therapist, 212
Fees, dilemmas regarding non-payment of, 115
Feminist and multicultural theories and past power relationships, 86
Feminist and multicultural therapies and systemic inequities, 90
Feminist and multicultural therapists, what they stress, 82
Findings, validity of generalizing, 8
Formal assessments and other consultation, 143
Formulation
clinical, 4
diagnostic, 4
treatment, 4
Function, and responsibilities, role, rules, 43
Functioning, hypothesis about client's, 126

G

Generalizing findings, validity of, 8
Gestalt therapy, 82
Goals
client's, 162–163
constraints on, 164–167
evaluating movement towards achieving agreed upon, 210
therapist's, 163–164
Goals with Nancy, developing, 167–169
Groups, knowledge of modal experiences of cultural, 50

H

Health and pathology, beliefs about, 26–33
what is pathological based on what is dysfunctional for individuals, 29
Helpful or constraining, diagnosis can be, 172
Helping, dilemmas in effective, 97–123
boundary dilemmas, 108–116
boundary issues related to managed care systems, 115–116
disclosure, 110–115
consultation and supervision, 121
fit of therapist and client, 98–107
social justice and cultural sensitivity, 116–121
clients' bias, 120–121
clients' internalized oppression, 118–120
discrimination related to therapist minority status, 120–121
Helping relationships, working alliance is foundation of all, 79
Homophobia, Juan's, 202
Hormonal imbalances, Juan and, 202
Hormone treatment began to work for Juan, 203
Human development, ecological model of, 2
Human experience, diversity of, 3
Humanistic and phenomenological therapists, 90
Hypothesis about client's functioning, 126

I

Identities, 40
Ideological differences, psychotherapy and, 104
Ideologies, 40

INDEX **263**

Images, disclosure and online social networking and online, 114
Individual, family, and social developmental challenges and attitudes, keeping abreast of, 209
Individuals
  client's story, 45
  in dominant status and sociostructural systems, 51
  relates differently to various characteristics of reference group, 49
  therapists who emphasize, 54–55
Inequities, feminist and multicultural therapies and systemic, 90
Inequity, privilege, and power, 22–24
Influences, ecological contexts and, 41t, 42f
Influences of personal worldviews, exploring, 17–37
  beliefs about health and pathology, 26–33
  importance of theoretical orientation, 17–19
  influences on ideas about change, 21–25
  introducing dimensions of change, 33–36
  social context of psychotherapy, 26–33
  understanding beliefs about change, 19–21
Information
  from client about client contexts, 131
  from clients about presenting problem, 130
  collateral, 142–143
  disclosing personal, 111
  organizing from various sources, 145
  primary sources of, 127–142
  secondary sources of, 142–145
  therapy and distinction between what clients present as, 182
Information, gathering and integrating, 125–149
  beginning process of case conceptualization, 125–145
  connecting information to theory, 146–148
  gathering information, 125–145
    formal assessments and other consultation, 143
    information from client's direct expressions, 127
    information from observations of and interactions with client, 137
    primary sources of information, 127–142
    secondary sources of information, 142–145
  moving towards conceptualization, 146–148
  organizing and integrating information (pre-conceptualization), 145–146
Information about client from observation, Nancy's therapist gathered, 152
Information and moving towards conceptualization with Nancy, integrating, 151–156
Information from client's, direct expressions, 127, 136
Information from clients, gathering, 137
Information from counselor-client interactions, 139
Information from interactions, challenge in evaluating and using, 140
Information from observations of and interactions with client, 137
Information necessary for conceptualization regarding Nancy, organizing, 153
Information obtained from clients, sharing, 143
Injury at work, physical, 134
Intake form, complete written, 128
Integrating information and moving towards conceptualization with Nancy, 151–156
Integrating information, gathering and, 125–149
  beginning process of case conceptualization, 125–145
  connecting information to theory, 146–148
  gathering information, 125–145
    formal assessments and other consultation, 143
    information from client's direct expressions, 127
    information from observations of and interactions with client, 137
    primary sources of information, 127–142
    secondary sources of information, 142–145
  moving towards conceptualization, 146–148
  organizing and integrating information (pre-conceptualization), 145–146
Integrationist orientation, therapists who adhered to, 111
Interact with others, wife's inability to, 144
Interactions
  challenge in evaluating and using information from, 140
  content of, 68
  feelings and, 83
  information from counselor-client, 139
Interactions with client, information from observations of and, 137
Interactive patterns, learning about and understanding, 44
Internalized oppression, characteristic of, 119
Internet, using, 49
Interventions
  in first few sessions of therapy, 127
  or conducting therapy, 8
Interviewing, motivational, 74
Intimacy, Nancy's need for relational, 185
Iterative conceptualization
  consultation, and adjunct treatments, 204–205
  and modifying diagnoses, 203–204
Iterative conceptualization and treatment planning, 179–207
  case of Juan, 194–203
    information from client from first five sessions, 194–196
    information from observations of client and counselor-client interactions, 196
    initial conceptualization and treatment, 197–199
    initial information, 194–197
    iterative conceptualization and treatment, 199–203
  continual conceptualization and integrating new information, 183–188
    revisiting organizing information for continual conceptualization, 184
    testing hypotheses from prior information and conceptualizations, 184–188
  continual conceptualization with Nancy, 181–183
  encountering surprises, 188–194
  interactions with systems of care, 206
  ongoing records and clinical notes, 205–206
  process of iterative conceptualization, 188–194
Iterative conceptualization from start to finish, 194–203
Iterative conceptualization from start to finish - case of Juan, 194–203
  information from client from first five sessions, 194–196
  information from observations of client and counselor-client interactions, 196
  initial conceptualization and treatment, 197–199
  initial information, 194–197
  iterative conceptualization and treatment, 199–203

J

Juan
  current lack of relationships, 200
  dating and sexual history, 201
  discussing high school relationships, 200
  homophobia, 202
  and hormonal imbalances, 202
  hormone treatment began to work for, 203

264 INDEX

Juan *(continued)*
  lack of sexual interest, 201
  re-conceptualizing, 201
Juan, case of - iterative conceptualization from start to finish, 194–203
  information from client from first five sessions, 194–196
  information from observations of client and counselor-client interactions, 196
  initial conceptualization and treatment, 197–199
  initial information, 194–197
  iterative conceptualization and treatment, 199–203
Juan's therapy and sleeping, tardiness, and new activities, 200
Justice, social - and cultural sensitivity, 116–121
  clients' bias, 120–121
  clients' internalized oppression, 118–120
  discrimination related to therapist minority status, 120–121

K

Knowledge
  about larger contextual issues, 134
  breath and depth of therapist's, 103
  of modal experiences of cultural groups, 50
  reflection, and experience, learning is a mix of, 3
  treatment changes over time in response to new, 179
Knowledge and experience, strategies for increasing, 48–49
Knowledge and skills, ways of teaching case conceptualization, 148
Knowledge or skill, novice therapists perceived as having insufficient, 100
Knowledge related to client's multiple contexts, skill and, 103

L

Language
  issues may also affect bilingual clients, 128
  some assumptions are conveyed directly in, 132
Learning is a mix of knowledge, reflection, and experience, 3
Literature within field, reading, 48
Living situation, Nancy's unhappiness with current, 182

M

Managed care and confidentiality, 116
Mandates, ethical, 105
Meanings, decoding, 145
Modal experiences of cultural groups, knowledge of, 50
Motivational interviewing, 74
Multicultural theories and past power relationships, feminist and, 86
Multicultural therapies and systemic inequities, feminist and, 90
Multicultural therapists, feminist and, what they stress, 82
Multiple contexts, skill and knowledge related to client's, 103
Multiple terminations, 225–226

N

Nancy
  anxiety and difficulty in managing practical issues, 181
  binge drinking and eating escalated, 189
  conceptualizing, 13–14
  developing goals with, 167–169
  and family therapy session, 152
  had heated arguments with parents, 183
  to have therapist's support and attention, 216
  identifying more long-term problems of, 182
  illustrates cyclical process of information, 187
  insisted that drinking was not a problem, 189
  integrating information and moving towards conceptualization with, 151–156
  introducing case of, 10–13
  investigated intense outpatient eating disorder program, 193
  is increasingly more comfortable in sessions, 186
  is more willing to share fears and reactions, 186
  need for relational intimacy, 185
  organizing information necessary for conceptualization regarding, 153
  participating more actively in sessions, 152
  unhappiness with current living situation, 182
  was going through honeymoon period, 182
Nancy, therapist initial treatment with, 181
Nancy, therapist modified conceptualization of, 185
Nancy, therapists initial thoughts about conceptualizing, 153
  challenges, 155–156
  how client's problems serve client, 155
  Nancy's perception of her distress, 153
  strengths and resources, 155
  therapist's conceptualization of contextual influences, 154–155
  therapist's views of Nancy's most pressing problems, 153–154
Nancy's mother threatened to stop therapy, 189
Nancy's therapist
  continues to focus more on affect and cognition, 187
  gathered information about client from observation, 152
Nancy's therapist set treatment goals and strategies, 186, 192–193
  binge behaviors on Nancy's academic achievement, 192
  continue to develop strong empathic and supportive relationship with Nancy, 186
  decrease Nancy's binge drinking and eating, 192
  deepen trust so that Nancy would become more easily and naturally forthcoming, 186
  develop Nancy's ability to care for herself, 186
  develop Nancy's ability to feel empowered to meet her own needs, 186
  develop Nancy's ability to take perspective of others, 186
  issue of continuing therapy Nancy's mother threatened to stop paying for, 193
Networking and online images, disclosure and online social, 114
No right or wrong answer, 83
Non-payment of fees, dilemmas regarding, 115
Novice therapist and theoretical orientation, 18
Novice therapists, 4
  affective change frequently most difficult for, 59
  are asking themselves if they are being effective, 211
  challenging, 214
  perceived as having insufficient knowledge or skill, 100

O

Observation, Nancy's therapist gathered information about client from, 152

Observations of and interactions with client, information from, 137
Online images, disclosure and online social networking and, 114
Online social networking and online images, disclosure and, 114
Operating rules, learning about and understanding, 44
Oppression, characteristic of internalized, 119
Organizational reasons, assigning diagnosis to your client for billing or, 173
Orientation
 clients inquiring about sexual, 112
 theoretical, 129
 therapists who adhered to integrationist, 111
Origin, family of, 134
Outcomes therapists evaluate, 212

P

Parents, Nancy had heated arguments with, 183
Past, therapists who emphasize role of, 62
Past power relationships, feminist and multicultural theories and, 86
Pathology, beliefs about health and, 26–33
Peer supervision or consultation, on-going, 121
People, foster relations with diverse, 49
People have, relationships that, 40
Personal differences in expectations, cultural or, 78
Personal information, disclosing, 111
Personal worldviews, exploring influences of, 17–37
 beliefs about health and pathology, 26–33
 importance of theoretical orientation, 17–19
 influences on ideas about change, 21–25
 introducing dimensions of change, 33–36
 social context of psychotherapy, 26–33
 understanding beliefs about change, 19–21
Phenomenological therapists, humanistic and, 90
Physical injury at work, 134
Planned termination, reasons for, 223
Populations, Empirically Supported Therapies (ESTs) require adaptation for differing, 8
Positive contextual factors, 167
Power, privilege, and inequity, 22–24
Power relationships, feminist and multicultural theories and past, 86
Preferences, dimensional, 129
Premature unplanned termination, 221–222
Present, therapists who emphasize, 62
Primary sources of information, 127–142
Privilege
 having more or less, 51
 power, and inequity, 22–24
Problems
 clients with unrelated, 105
 deciding that something is, 58
 diagnosis validates that there is a, 172
 Nancy insisted that drinking was not a, 189
Problems of Nancy, identifying more long-term, 182
Process
 communication rarely captured in single statement, 89
 conceptualization is relational, 7
 involves formulating response indicating understanding, 87
 systems therapy very consistent with focusing on, 90
 therapist's awareness of, 89
Process approach, less structured, 70
Process of change, understanding, 2
Process of continual conceptualization, difficult to describe, 180

Processes, learning about and understanding communication, 44
Progress, therapists incorporate clients' own evaluations of, 211
Psychodynamic theories, 63, 86, 156
Psychodynamic therapists, 90
Psychological experience, contexts affecting development of, 53
Psychotherapy
 and ideological differences, 104
 purpose of, 19
 research on, 7
 social context of, 26–33

Q

Questions, skills in conceptualization and addressing, 1

R

Real and unreal relationships, therapists focus on both, 83–84
Re-conceptualizing Juan, 201
Referral, communicating reasons for, 107
Referral context, 126
Referral to another therapist, making, 107
Referring client for formal assessment, 144
Reflection, and experience, learning is a mix of knowledge, 3
Regional community events and resources, searching for, 49
Relational dimensions, preferences on, 142
Relational endeavor, therapy is inherently, 6
Relational intimacy, Nancy's need for, 185
Relational norms, expectations, values and cultural experiences, 78
Relational process, conceptualization is, 7
Relational skills, therapists frequently focus on developing, 1
Relational style as therapist, conceptualizing involves, 67
Relational systems, therapist who emphasize, 55
Relational theory, 156
Relations with diverse people, foster, 49
Relationship roles, learning about and understanding, 44
Relationships
 centrality of, 81
 dynamics in therapeutic, 85
 feminist and multicultural theories and past power, 86
 that people have, 40
 therapeutic, 72
 with therapists and change, 81
 therapists focus on both real and unreal, 83–84
 therapists who emphasize, 55
 therapy is particular kind of, 108
 working alliance is foundation of all helping, 79
Relationships, conceptualizing therapeutic, 67–96
 pulling it all together, 91–95
 relationships and location of knowledge, 90–91
 role of relationship in contributing to change, 76–90
  process emphasis, 87–90
  process emphasis dimension to established theories, 89
  real unreal relationship, 82–87
  real-unreal relationship dimension to establish theories, 86
  therapeutic relationship, 79–82

therapeutic relationship dimension to established theories, 81–82
working alliance, 77–79
understanding relational style, 68–76
Relaxation training, biofeedback and, 133
Research on psychotherapy, 7
Resources, searching for regional community events and, 49
Responses will affect therapy, how various, 120
Responsibilities, role, rules, function, and, 43
Role, rules, function, and responsibilities, 43
Roles, learning about and understanding relationship, 44
Rules
  function, and responsibilities, role,, 43
  learning about and understanding operating, 44

S

Secondary sources of information, 142–145
Self-disclosure, therapists need to consider, 86
Self-evaluation as therapist, feelings and thoughts guiding, 212
Self-examination, need for continual, 141
Sessions
  effectiveness within, 214
  evaluation of given, 214
  Nancy and family therapy, 152
  Nancy is increasingly more comfortable in, 186
  Nancy participating more actively in, 152
Sessions of therapy, interventions in first few, 127
Sexual orientation, clients inquiring about, 112
Skills
  in conceptualization and addressing questions, 1
  and knowledge related to client's multiple contexts, 103
  novice therapists perceived as having insufficient knowledge or, 100
  therapists frequently focus on developing relational, 1
  ways of teaching case conceptualization knowledge and, 148
Social context of psychotherapy, 26–33
Social developmental challenges and attitudes, keeping abreast of individual, family, and, 209
Social justice and cultural sensitivity, 116–121
  clients' bias, 120–121
  clients' internalized oppression, 118–120
  discrimination related to therapist minority status, 120–121
Social networking and online images, disclosure and online, 114
Social statuses, addressing similarities or differences in, 106
Sociocultural aspect, 46
Sociocultural systems
  have complex intersections, 52
  as related to power and privilege, 50
  therapists who emphasize, 56
Sociocultural/sociostructural system, therapists who emphasize, 55
Sociostructural aspect, 50
Sociostructural system, ethnic culture is, 51
Sources of information
  primary, 127–142
  secondary, 142–145
Spare the rod and spoil the child, 135
Story, individual client's, 45

Story about client, conceptualization becomes embedded in ongoing, 185
Struggles, therapists and clients, 141
Supervision, consultation and, 121
Supervision or consultation, on-going peer, 121
Support and attention, Nancy to have therapist's, 216
Systemic inequities, feminist and multicultural therapies and, 90
Systems therapy very consistent with focusing on process, 90

T

Task Force on Evidence Based Practice from American Psychological Association (APA), 9
Terminations
  are frequently difficult for therapists as well as clients, 220
  continuous evaluation and, 209–231
  multiple, 225–226
  nature of, 220
  planned, 223–225
  premature unplanned, 221–222
Theoretical orientations
  case conceptualization founded upon, 147
  case conceptualization is applied, 6
  conceptualization is application of, 5–6
  and conceptualization not the same, 5
  and dimensional preferences, 129
  is foundation for conceptualization, 6
  making choices about, 19
  novice therapist and, 18
  relation of dimensions to traditional, 36t
Theoretical orientations and dimensional preferences, therapist influenced by, 187
Theories
  attachment, 156
  change dimension and established, 61
  cognitive-behavioral, 156
  context emphasis on dimension and established, 57
  established them, 75
  family systems, 57, 156
  past or present emphasis dimension and established, 63
  psychodynamic, 63, 86, 156
  relational, 156
Theories and past power relationships, feminist and multicultural, 86
Therapeutic interventions contribute to positive change, 182
Therapeutic relationships
  dynamics in, 85
  special kind of relationship, 72
Therapeutic relationships, conceptualizing, 67–96
  pulling it all together, 91–95
  relationships and location of knowledge, 90–91
  role of relationship in contributing to change, 76–90
    process emphasis, 87–90
    process emphasis dimension to established theories, 89
    real unreal relationship, 82–87
    real-unreal relationship dimension to establish theories, 86
    therapeutic relationship, 79–82
    therapeutic relationship dimension to established theories, 81–82
    working alliance, 77–79
  understanding relational style, 68–76

Therapies and systemic inequities, feminist and multicultural, 90
Therapist
  initial treatment with Nancy, 181
  modified conceptualization of Nancy, 185
  was not on Nancy's family health plan, 189
Therapist and change, relationship with, 81
Therapist and client, fit of, 98–107
Therapist does or says, client not happy with something, 215
Therapist gathered information about client from observation, Nancy's, 152
Therapist set treatment goals and strategies, Nancy's, 186
Therapists
  affective change frequently most difficult for novice, 59
  awareness of processes, 89
  boredom, 140
  challenging novice, 214
  client characteristics contributing to comfort or discomfort of, 100
  client discomfort or even criticism of, 215
  and clients struggles, 141
  conceptualizing involves relational style as, 67
  differences between discomfort and criticism of, 216
  directive, 69
  disclosures about themselves, 111
  faced with dilemmas about disclosure, 113–114
  feelings and thoughts guiding self-evaluation as, 212
  focus on both real and unreal relationships, 83–84
  frequently focus on developing relational skills, 1
  goals, 163–164
  humanistic and phenomenological, 90
  incorporate clients' own evaluations of progress, 211
  influenced by theoretical orientations and dimensional preferences, 187
  intuitively modify their approach to clients, 140
  making referral to other, 107
  need to consider self-disclosure, 86
  not liking clients, 107
  novice, 4, 100
  psychodynamic, 90
  who adhered to integrationist orientation, 111
  who emphasize individuals, 54–55
  who emphasize present, 62
  who emphasize relational systems, 55
  who emphasize relationships, 55
  who emphasize role of past, 62
  who emphasize sociocultural systems, 56
  who emphasize sociocultural/sociostructural system, 55
Therapists as well as clients, terminations are frequently difficult for, 220
Therapists evaluate, outcomes, 212
Therapists initial thoughts about conceptualizing Nancy, 153
Therapist's knowledge, breath and depth of, 103
Therapist's support and attention, Nancy to have, 216
Therapy
  derailing, 213
  dilemmas that may emerge in, 97
  disclosing something from very beginning of, 113
  and distinction between what clients presents as information, 182
  emotion-focused, 82
  gestalt, 82
  how various responses will affect, 120
  interventions in first few sessions of, 127
  interventions or conducting, 8
  is inherently relational endeavor, 6
  is particular kind of relationship, 108
  Nancy's mother threatened to stop, 189
  preference for facilitating change in, 34t
  reasons for seeking, 105
  systems, 90
  trainers, 4
  unplanned termination in beginning phases of, 222
  working alliance in, 77
Therapy session, Nancy and family, 152
Thoughts guiding self-evaluation as therapist, feelings and, 212
Trainee's fear, 101
Trainers, therapy, 4
Training, biofeedback and relaxation, 133
Transferential elements, 84
Treatment, cyclical process of conceptualization and, 188
Treatment changes over time in response to new knowledge, 179
Treatment formulation, 4
Treatment goals and strategies, Nancy's therapist set, 186
Treatment plan, case conceptualization and developing effective, 147
Treatment planning, and diagnosis, conceptualization, 151–177
  choosing strategies for treatment goals, 169–171
  conceptualization and treatment planning, 158–169
    developing goals, 162–169
      therapists' intervention preferences, 158–160
      treatment planning, journey, not a destination, 160–162
  diagnosis, 171–175
  formalizing conceptualizations and treatment plans, 175–176
  information gathering with early planning, 156–158
    initial treatment planning during information gathering with Nancy, 157–158
  moving towards conceptualization with Nancy, 151–156
Treatment planning, iterative conceptualization and, 179–207
  case of Juan, 194–203
    information from client from first five sessions, 194–196
    information from observations of client and counselor-client interactions, 196
    initial conceptualization and treatment, 197–199
    initial information, 194–197
    iterative conceptualization and treatment, 199–203
  continual conceptualization and integrating new information, 183–188
    revisiting organizing information for continual conceptualization, 184
    testing hypotheses from prior information and conceptualizations, 184–188
  continual conceptualization with Nancy, 181–183
  encountering surprises, 188–194
  interactions with systems of care, 206
  iterative conceptualization
    consultation, and adjunct treatments, 204–205
    and modifying diagnoses, 203–204
  iterative conceptualization from start to finish, case of Juan
    information from client from first five sessions, 194–196

information from observations of client and counselor-client interactions, 196
initial conceptualization and treatment, 197–199
initial information, 194–197
iterative conceptualization and treatment, 199–203
ongoing records and clinical notes, 205–206
process of iterative conceptualization, 188–194
Treatment with Nancy, therapist initial, 181

U

Understanding
 making choices about what is central to, 147
 process involves formulating response indicating, 87
Unplanned termination, 221–222
Unreal relationships, therapists focus on both real and, 83–84
Unrelated problems, clients with, 105

V

Validity of generalizing findings, 8
Values and beliefs being absolute, 104
Variables, matching on, 106
Verbalizations from client, 135

W

Wife's inability to interact with others, 144
Work, physical injury at, 134
Working alliance, 77, 79
Worldviews, exploring influences of personal, 17–37
 beliefs about health and pathology, 26–33
  criteria for evaluating health and pathology, 27–30
  influences on conceptualization of health and pathology, 30–33
  what is pathological according to accepted social standards, 28–29
  what is pathological according to authorities, 29
  what is pathological according to modal or average experiences, 28
  what is pathological based on what is dysfunctional for individuals, 29
 importance of theoretical orientation, 17–19
 influences on ideas about change, 21–25
  beliefs about people and worldview, 21–22
  cultural influences on understanding clients, 21–22
  power, privilege, and inequity, 22–24
  power and privilege related to role of therapist, 24–25
 introducing dimensions of change, 33–36
 social context of psychotherapy, 26–33
  criteria for evaluating health and pathology, 27–30
  cultural influences on understanding health and pathology, 31–32
  individual functional influences on conceptualizing health and pathology, 32–33
  influences on conceptualization of health and pathology, 30–33
  what is pathological according to accepted social standards, 28–29
  what is pathological according to authorities, 29
  what is pathological according to modal or average experiences, 28
  what is pathological based on what is dysfunctional for individuals, 29
 understanding beliefs about change, 19–21
Written intake form, completing, 128